Japan Science & Technology Outlook

Japan Science & Technology Outlook

Based on *Kagakugijutsu Hakusho*, a white paper of the Science and Technology Agency

FUJI CORPORATION
Busicen Bldg., 5-29-7, Jingu-mae, Shibuya-ku, Tokyo 150, Japan
Phone: (03) 409-6291 Telex: 02425496 FUJICO J

First published in Japanese as *Kagakugijutsu Hakusho 1982* by the Science and Technology Agency, and printed by the Printing Bureau, Ministry of Finance in 1983.

English edition published by Fuji Corporation, Busicen Bldg., 5-29-7, Jingumae, Shibuya-ku, Tokyo 150, Japan

Printed in Japan

ISBN4-938329-11-5 C3050

PREFACE

Japan Science and Technology Outlook is based on the latest white paper of the Science and Technology Agency. To make it more appropriate and comprehensible to foreign readers, a number of sections of little relevance have been omitted, repetitious passages eliminated, necessary explanations added, and the order of certain sections revised. All yen values were converted into U.S. dollars, at the rate of 250 yen to the dollar, and then rounded off. The original text, entitled *Kagakugijutsu Hakusho*, is available from the Science and Technology Agency.

We would like to thank the Science and Technology Agency for their cooperation in publishing the book, and Mr. Robert Newkirk for his assistance in editing the English text.

August 1983 Fuji Corporation

CONTENTS

INTRODUCTION

Science and technology have made great contributions to social and economic progress, and in recent years, as awareness of these benefits has heightened internationally, science has increasingly come to be viewed hopefully as the answer to many current problems, such as economic stagnation or the limitations of natural resources and energy.

Japan, both socially and economically, developed rapidly after the end of World War II, in large part due to progress in science and technology. In fact, in fields such as electronics and industrial robots, Japan has advanced to the point where it is now leading the world. Nevertheless, it can still be said that development of Japanese science and technology has depended heavily upon technologies introduced from abroad.

In order for Japan, now one of the world's leading nations in science and technology, to solve the many problems it faces both domestically and internationally, it needs to overcome its dependency upon technologies introduced from abroad, cut a new path for itself in the field of science and technology, and further even more creative research and the development of innovative technologies.

CHAPTER 1

SCIENCE AND TECHNOLOGY IN JAPAN

1. PROGRESS IN TECHNOLOGY

Since the end of World War II, the Japanese economy has grown steadily, to the point where the country now possesses the second greatest economic strength in the free world. Among the many factors behind this unprecedented growth, a major one has been progress in technological innovation.

Because of the technological vacuum during World War II, Japan, finding itself at the close of the war far behind the technological level of Europe and the U. S., imported proven and promising technologies from those countries, and then improved and adapted those technologies to increase productivity and thereby acquire internationally competitive power.

Figure 1-1 shows quantitatively the extent to which technological progress contributes to production increases in manufacturing. According to the figure, production in manufacturing industries increased nearly 12-fold during the period from 1950 to 1979, and technological progress accounts for approximately 30 percent of this growth, showing that it fulfills a considerably significant role. The percentage is extremely high in the machinery industry, a typical processing-and-assembly type industry.

Next, summarizing the economic results of this progress, Figure 1-2 shows changes in total annual shipments made in several industries from 1960 to 1979, indicating the significant growth achieved in the heavy and chemical industries between 1965 and 1970.

This increase is even more apparent in Japan's export structure. Figure 1-3, showing the ratio of exports by industry from 1960 to 1975, points up the conspicuous growth in the share of heavy and chemical industries, the major cause of which lies in the growth of the processing-and-assembly type industries.

Fig. 1-1. Contribution of Technology to Production

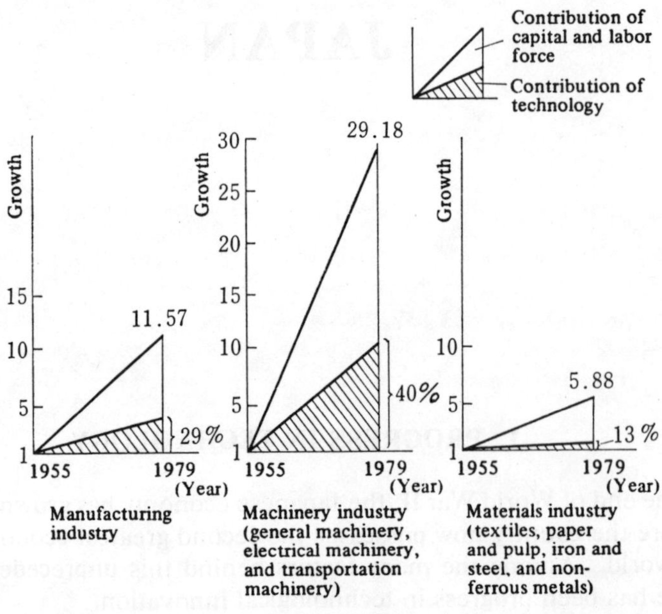

Note: The peak figures on the above graphs represent the degree of growth for 1979, where the degree for 1955 is represented as one, while the percentages represent the contribution of technological progress.

Source: The Economic Planning Agency

Fig. 1-2. Changes in Ratio of Shipment Values by Industry

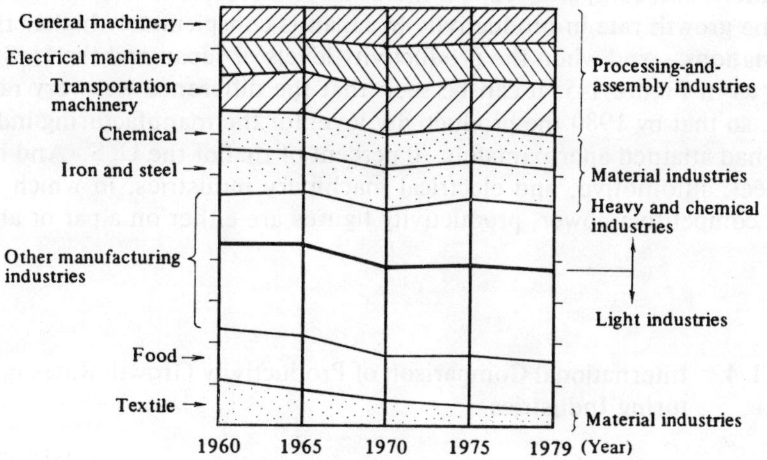

Source: "Industrial Statistics Tables," the Ministry of International Trade and Industry

Fig. 1-3. Changes in Ratio of Export Values by Industry

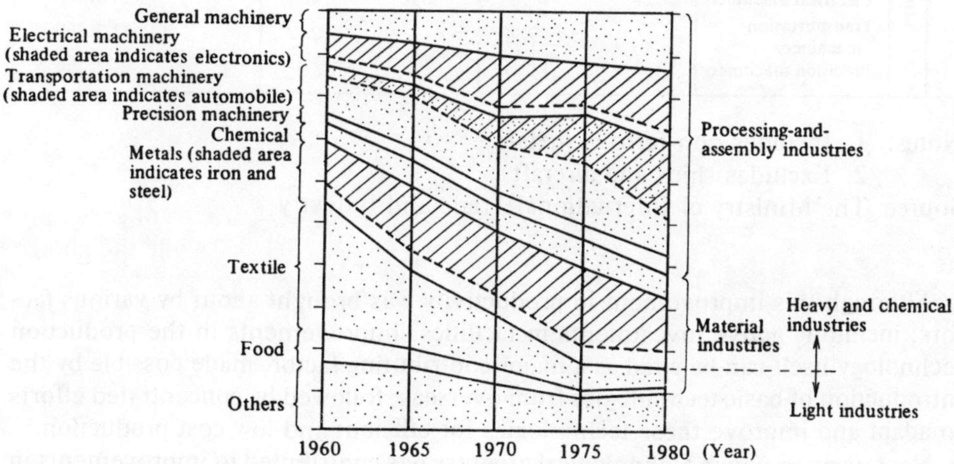

Source: The Ministry of International Trade and Industry

As seen in these two figures, the share of processing-and-assembly type industries, in which it is relatively easy to enhance the added value of products, has risen both in production and exports.

Furthermore, comparing Japan, the U. S., and West Germany, Table 1-4 shows that, the growth rate in productivity in Japan is conspicuously higher than in the other nations. And when the productivity levels of Japan and the U. S. are compared, as in Figure 1-5, it can be seen that the difference was very rapidly narrowed, so that by 1980 the productivity level for the manufacturing industry as a whole had attained approximately 70 percent of that of the U. S. And in the iron and steel, automotive, and electrical machinery industries, in which Japan has strong competitive power, productivity figures are either on a par or above U. S. levels.

Table 1-4. International Comparison of Productivity Growth Rates in Manufacturing Industries

(Unit: percentage)

		Japan (average for 1960–1979)	United States (average for 1960–1979)	West Germany (average for 1960–1979)
Manufacturing industries		8.2	3.3	5.5
Industry	Textile	8.6	4.1	6.8
	Chemical	10.4	5.5	7.5
	Iron and steel	9.8	2.3[1]	4.2
	General machinery	9.9	2.8	2.8
	Electrical machinery	10.9	4.5	6.5
	Transportation machinery	9.4	2.2	3.9[2]
	Precision machinery	11.5	—	4.2

Notes: 1. Includes non-ferrous metals
2. Excludes ships and aircraft
Source: The Ministry of International Trade and Industry

Although this improvement in productivity was brought about by various factors, including active investments in facilities, improvements in the production technology itself can be cited as a major contributing factor, made possible by the introduction of basic technologies from overseas, followed by concentrated efforts to adapt and improve these technologies for efficient and low cost production.

Next, ways in which technological progress has contributed to improvements in productivity and competitive power will be discussed for the iron and steel industry, the automotive industry, and the semi-conductor industry.

Fig. 1-5. Comparison of Productivity Levels in Japan and U. S. Manufacturing Industries

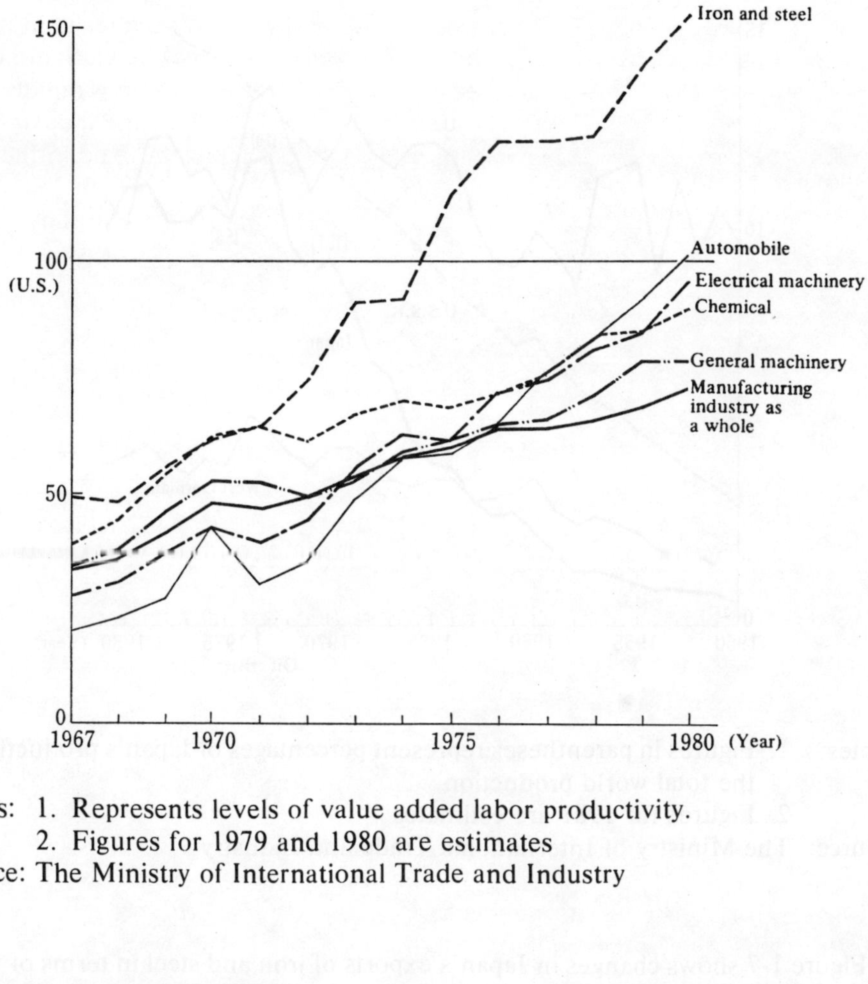

Notes: 1. Represents levels of value added labor productivity.
2. Figures for 1979 and 1980 are estimates

Source: The Ministry of International Trade and Industry

(1) The Iron and Steel Industry

The iron and steel industry of Japan currently operates at the highest level in the world in terms of facilities and equipment, manufacturing costs, quality, and technological capabilities, and is said to be preeminent in terms of international competitive power.

Figure 1-6 shows changes in the volumes of production of crude steel in the leading nations of the world. As can be seen, the growth in the volume of production in Japan, a major factor behind the nation's high economic growth rate, is conspicuous, outstripping West Germany in 1964, and then, in 1980, surpassing even the U. S.

Fig. 1-6. Changes in Crude Steel Production

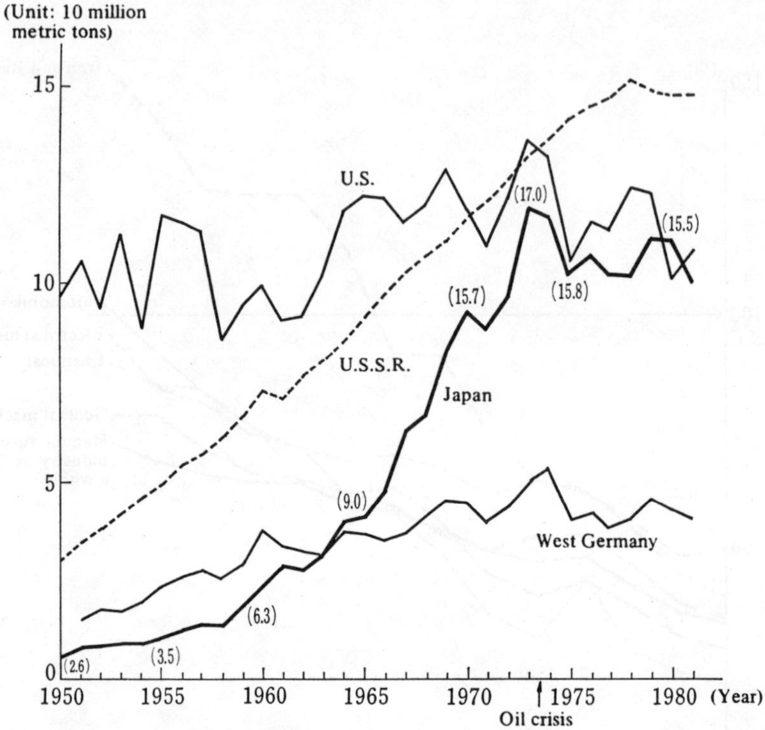

Notes: 1. Figures in parentheses represent percentages of Japan's production in the total world production
2. Figures for 1981 are estimates

Source: The Ministry of International Trade and Industry

Figure 1-7 shows changes in Japan's exports of iron and steel in terms of value and volume, which increased rapidly during the ten year period from 1965 to 1975. Following the international economic stagnation caused by the oil crisis, although the volume of exports decreased, their value continued to increase rapidly. This is due to the fact that while growth in the value of exports prior to the oil crisis depended on quantitative increases in low-priced mass-produced products, in later years product diversification and qualitative improvements such as a shift towards high added-value products have played a significant role.

While there are other factors accounting for the rapid growth in the iron and steel industry in Japan, improvement in technological standards is one of the major factors.

Fig. 1-7. Changes in Japan's Iron and Steel Exports

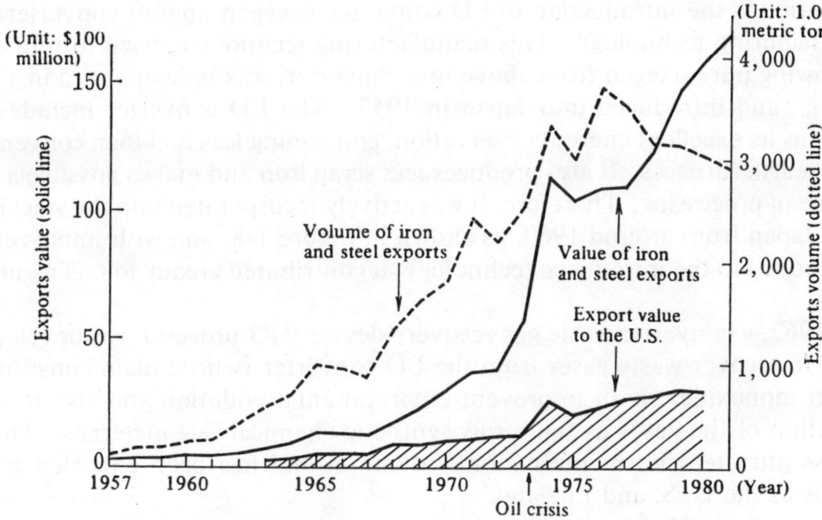

Source: The Ministry of International Trade and Industry

In a questionnaire sent to private enterprises in Japan in January, 1982, more than 70 percent of respondents replied that technological standards in the iron and steel industry in Japan were more advanced than in the U. S. The results of the poll also showed that Japanese corporations are more or less on a par with those in Europe and the U.S. in their ability to create new technologies and products.

The iron and steel industry of Japan, embarking upon reconstruction after nearly total destruction at the end of the war, owes much to the national policy of the "priority production systems," involving such things as preferential allocation of energy sources and priority financing.

It also owes a great deal to the advanced technologies from Europe and the U.S. that were introduced one after the other, starting with technological guidance from the U.S. during the 1950's.

The First Rationalization Plan, consisting mainly of the modernization of rolling equipment, began in 1951, and the latest facilities, equipment, and technologies were actively introduced. In that year, strip mill technology, in which long belt-form steel plate is manufactured from steel ingots, was introduced from the U.S., resulting in major productivity gains. Subsequently, domestic production of strip mill technology progressed, and, by providing high quality steel plate, contributed greatly to quality improvements in products such as motor vehicles and home electrical appliances.

The Second Rationalization Plan began in 1956, aimed at installing new integrated facilities to carry out all processes from the manufacturing of pig iron to the rolling process, and thereby increase production. Particularly noteworthy during this period is the introduction of LD converter (oxygen updraft converter) steel manufacturing technology. This manufacturing technology, used to refine steel by blowing pure oxygen from above in a converter, was industrialized in 1953 in Austria, and introduced into Japan in 1957. The LD converter includes such merits as its excellent energy conservation, consuming less fuel than conventional open hearth furnaces. It also produces less scrap iron and makes possible a larger volume of processing. Therefore, it was actively incorporated into the steel industry in Japan from around 1960, as shown in Figure 1-8, and with improvements being added to the introduced technology, it contributed greatly to raising productivity.

In 1962, a converter waste gas recovery device (OG process) was developed in Japan to recover waste gases from the LD converter (whose main constituent is carbon monoxide), both to prevent environmental pollution and also to enable utilization of the gases as fuel or as synthetic chemical raw materials. This OG process attracted much international attention and has been exported to such nations as the U. S. and England.

From around 1960, the construction of large-scale, high-performance integrated steel mills at new locations got underway actively, and iron and steel production grew dramatically. Also during this period, to more effectively use such

Fig. 1-8. Changes in Steel Production Processes in Japan

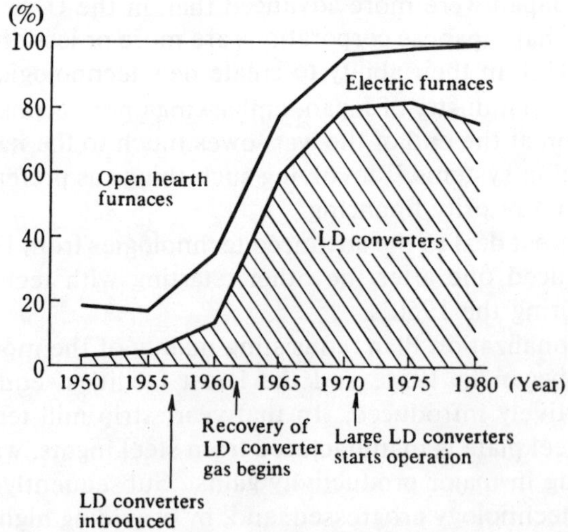

Source: The Committee on Iron and Steel Statistics

materials as iron ore and raw stock coal, for which Japan's degree of dependence is very high, and to achieve higher efficiency in operating blast furnaces, such measures as pre-treatment of ore, high-pressure operation, blowing-in of fuels such as heavy oil or tar, oxygen enrichment, and forced feeding of high-temperature humidity-controlled air were added one after the other. As a result of these efforts, as shown in Figure 1-9, the pig iron production ratio (pig iron produced per day per cubic meter of blast furnace internal capacity) increased, while the fuel ratio (fuel consumption required for the production of one ton of pig iron) decreased, both ratios having in recent years attained the highest standards in the world.

Fig. 1-9. Changes in Blast Furnace Productivity

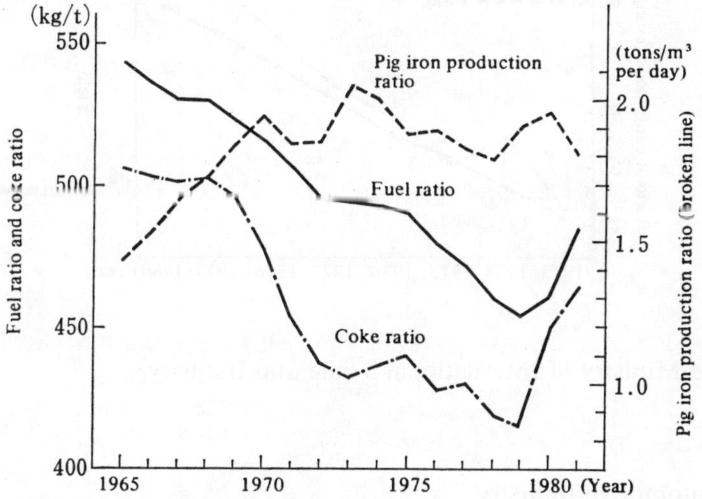

Source: The Japan Iron and Steel Federation

Also occupying a significant place in iron and steel technology is the introduction of the continuous casting process, a method for fabricating steel stock directly from molten steel refined in furnaces such as converters in a continuous fashion. Introduction of this process had already taken place in 1955, but at that time, with problems still to be solved in such areas as productivity, product quality, equipment and facilities costs, it took some time for the process to be placed into widespread use. However, this technology improved the yield (the ratio of finished steel products to steel ingots used), mechanization of the work, and conservation of manpower, and in addition an energy saving process. Therefore, research and development to solve these problems was carried out, so that, as shown in Figure 1-10, starting around 1970 the process rapidly went into widespread use, and the world's first integrated steel mill employing the continuous casting process was built in Japan in 1972. By 1981, continuous casting was used in approximately 60 percent of the industry in Japan, a high level compared to other advanced nations, such as the figures of 20 percent for the U. S., 27 percent for England, 41 percent for France, and 45 percent for West Germany.

Fig. 1-10. Continuous Casting and Yield

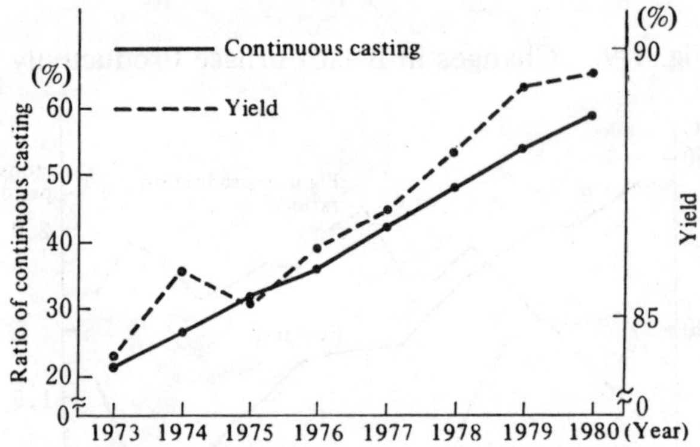

Source: The Ministry of International Trade and Industry

(2) The Automotive Industry

The automotive industry of Japan, developing very rapidy after World War II, is today equipped with the world's highest technological standards and competitive power in the field of small cars. Examining growth in the automotive industry in terms of production and trade, Figure 1-11 shows that, first of all, by the mid-1960's the number of motor vehicles manufactured in Japan surpassed the figures for Europe, while by 1980, it surpassed those for the U. S. Next, Figure 1-12, concerning the number of motor vehicles exported, shows that since the mid-1970's Japan has been the number one exporter in the world.

In a questionnaire sent in 1982 to private enterprises in Japan, approximately 40 percent of the respondents replied that the standards of Japanese automotive corporations are somewhat superior to those in Europe and the U. S., and, combined with those replies that said the respective national standards are on a par, the figure rises to approximately 90 percent. However, in standards of technological development, 40 percent replied that Japanese levels are below those of Euro-

Fig. 1-11. Automobile Production in Major Countries

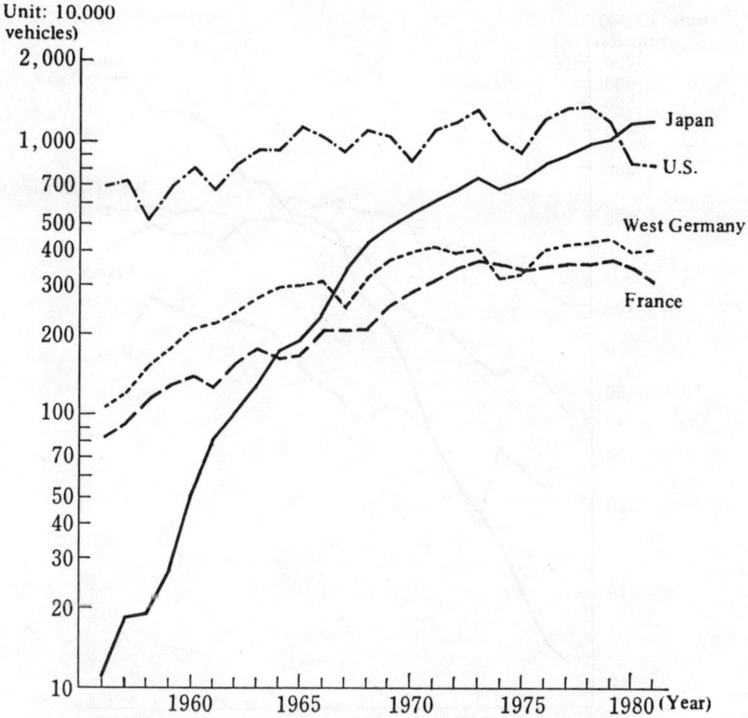

Source: The Japan Automobile Manufacturers' Association

pean and U. S. corporations. In other words, these comments reflect the fact that although great efforts in research and development work based on technologies introduced from abroad have resulted in technological standards surpassing even those of European and U. S. corporations, independent technological development still remains a problem.

Since the basic technologies for mass production of motor vehicles were well established in Europe and the U. S. even before an automotive industry came into being in Japan, virtually all automotive technologies in Japan have been introduced from abroad, then adapted and improved upon. In particular, technological standards in Japan improved greatly through technological tie-ups with foreign corporations concluded after World War II, or through the introduction of the very latest foreign equipment and facilities.

But because automotive technology depends on integration of many individual technologies, much independent technological development work has been carried out to meet conditions in Japan or the needs of the market, and these efforts

Figure 1-12. Autmobile Exports of Major Countries

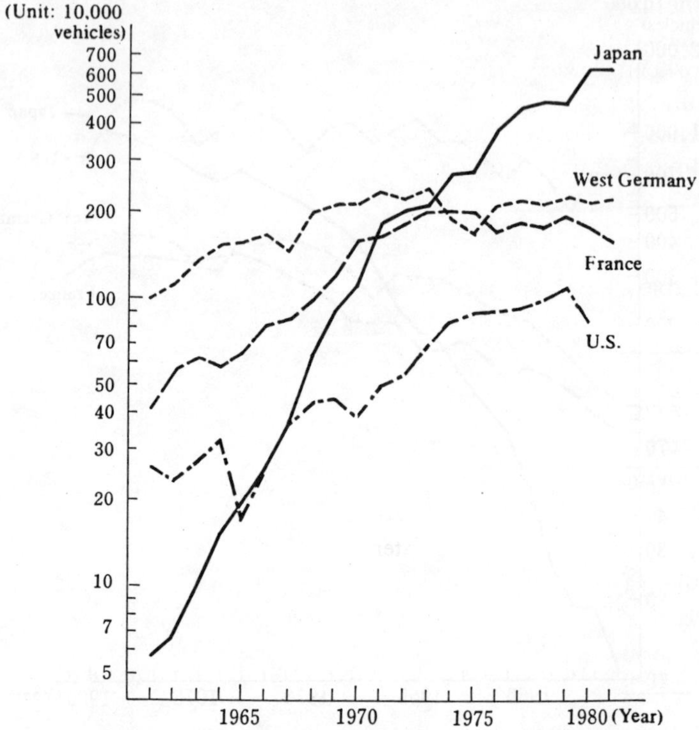

Source: The Japan Automobile Manufacturers' Association

have served as an important factor in the strong competitive power of Japanese motor vehicles today.

In production technology, because the Japanese automotive industry has long faced intense competition in both domestic and international markets, energetic efforts were made to improve productivity through such means as technological development, investments in equipment and facilities for new technologies, and improvements in production methods.

Also, because Japan's degree of dependence on foreign nations for energy is very high and its energy costs are relatively high, following the onset of the oil crisis, implementation of energy conservation measures has repeatedly been given priority in production processes. As a result, energy consumption at motor vehicle production plants in Japan is gradually decreasing, as shown in Figure 1-14.

Automation of production processes is also being pushed forward. In particular, in recent years, rapid development in electronic technologies is being put to practical use to raise production. For example, computer-assisted designing has made possible the production of highly functional and dependable bodies and chasses, while automation of production processes with electronic devices and equipment has contributed in a major way to improvements in productivity, attracting world attention as a major factor behind the strong competitive power of Japan's automotive industry. In particular, industrial robots are being introduced more energetically than in other industries, and are used extensively in such process as welding, painting, and assembly.

Table 1-13. Examples of Energy Conservation Measures

[Process]	[Example of measure]
Casting	Thermal insulation for melting furnaces
Forging	Improvement of heating efficiency
Sintering	Thermal insulation for furnace
Machining	Optimizing of mechanical capabilities
Press work and welding	Intermittent operation of hydraulic units
Painting	Recovery of waste heat from drying ovens
Plastic molding	Optimizing of heating temperatures
Mechanical fitting and assembly	Prevention of air leaks in tools

Source: The Japan Automobile Manufacturers' Association

Fig. 1-14. Energy Used in Automotive Production

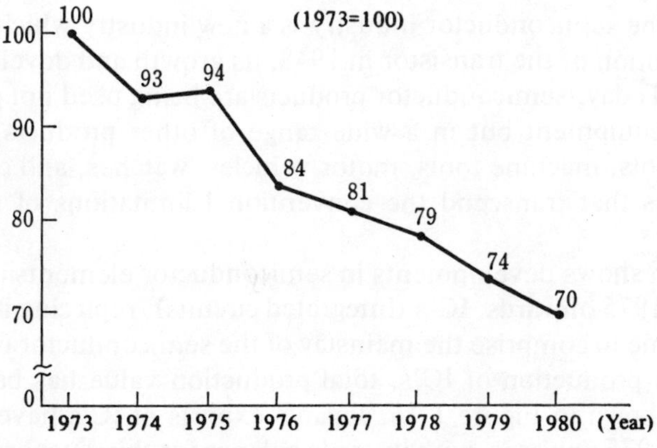

Source: The Japan Automobile Manufacturers' Association

At the same time, in regard to technology which affects product quality and performance, in recent years technologies to meet demands for low fuel consumption and low pollution gained much importance.

The importance of energy conservation measures heightened, triggered off by the petroleum crisis, and calls were made for reductions in the size of motor vehicles and also improvements in the fuel economy of the vehicles themselves. The technology involved in reducing motor vehicle size is not simply one of reducing dimensions. In order to materialize in small vehicle bodies the same functions and performance as large vehicles, rational design technologies and manufacturing technologies that minimize waste are necessary, while cost reductions must be pursued in a much more severe manner than with large vehicles. In Japan, efforts were made from an early date to develop a small motor vehicle with good fuel economy, because of such things as national energy supplies and road conditions, and thanks to an accumulated experience in these areas, the high standards of small vehicle technology have made it possible for Japan to reach a position of predominance in the recently expanding worldwide market for small vehicles.

Introduction of electronics in motor vehicles was triggered off by the need for low pollution vehicles, and efforts were made to adapt various types of sensors and microprocessors for engine control.

This trend, made possible by developments in solid state technologies such as microprocessors and pushed on by increasing demands for more sophisticated and diversified performance, is proceeding at a very fast rate to improve operating performance, fuel consumption, exhaust emission control, safety, and comfort.

Although incorporation of electronics began in the U. S., in Japan, as a result of having to meet extremely severe standards in exhaust emission control, plus energetic research and development by the automotive and electronic industries, automotive electronics technology has advanced very rapidly, and today maintains the highest standards in the world.

(3) The Semiconductor Industry

Although the semiconductor industry is a new industry which came into being with the invention of the transistor in 1948, its growth and development has been spectacular. Today, semiconductor products are being used not only in electronic devices and equipment but in a wide range of other products as well, such as industrial robots, machine tools, motor vehicles, watches, and cameras, bringing about changes that transcend the conventional limitations of mechanical technology.

Figure 1-15 shows developments in semiconductor elements in Japan in recent years. From 1975 onwards, IC's (integrated circuits), replacing individual transistors, have come to comprise the mainstay of the semiconductor industry, and with the growth in production of IC's, total production value has been growing very rapidly. As shown in Figure 1-16, Japan's exports of IC's have also grown very rapidly from 1975 onwards, with its trade balance for this category moving into the black in 1979 in terms of trade with the entire world, and in 1980 in terms of trade with the U. S. Although the semiconductor industry of Japan has grown very rapidly, nevertheless, in the world IC market in 1980, U. S.-affiliated corporations took a 68 percent share to Japan's 26 percent, giving the U. S. continued great predominance.

Fig. 1-15. Changes in Production of Semiconductors

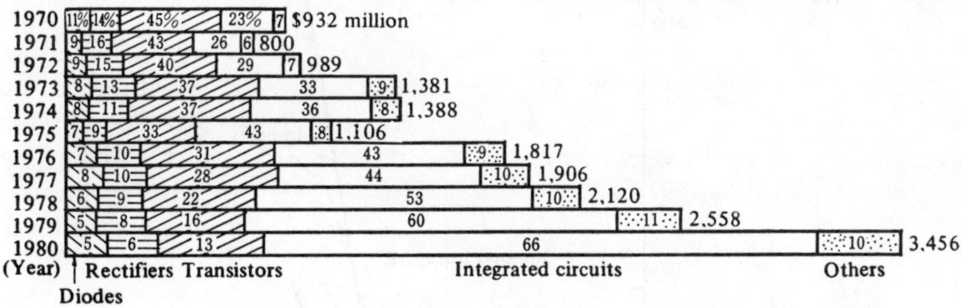

Source: *Electronic Industry Annual*, Dempa Publications, Inc.

Figure 1-17 compares the production structure of IC's in Japan and the U. S. according to type. In Japan, MOS (metal-oxide semiconductor) types, at over 50 percent of production, comprise the mainstay. Although its speed is somewhat slow, its power consumption is low, and structurally it allows with relative ease a high degree of integration and is suited to volume production, making it ideal for consumer applications. As a result of active research and development on the MOS type from an early date, Japan has acquired international competitive power in the fields of memory IC's and various other types of IC's for consumer use that are suited to volume production, so that by 1980, 50 percent of the MOS memory IC's manufactured in Japan were exported, taking a share in the world market as high as 20 percent. The bipolar linear type used in consumer appliances, especially stereo systems, also accounts for a considerable share of Japan's total. In this way, in Japan, IC's that are suited for consumer applications and feature good productivity occupy the main part of production.

In the U. S., while here again the ratio of the MOS type is high, compared to Japan the proportion of the bipolar digital type is extremely high. This type was developed earlier than the MOS type, and because it has the advantage of functioning at a high speed, active research and development work supported by military, space, or industrial demands has placed the U. S. in position of predominance. Also, the U. S. has strong competitive power in such fields as logic IC's not suitable for volume production, or custom IC's (special custom specification IC's) that require microcomputer or sophisticated system designing.

Fig. 1-16. Changes in IC Trade

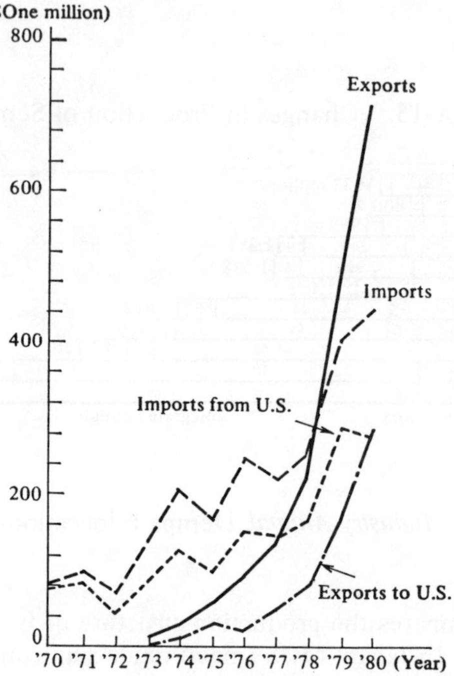

Source: The Economic Research Institute, Japan Society for the Promotion of Machine Industry

Virtually all inventions and developmental breakthroughs in semiconductor elements have taken place in the U. S., including both the transistor in 1948, which marks the start of semiconductors, and the IC at the end of the 1950's. In 1960, the MOS type transistor was developed and adapted for use in IC's, which started moving on the path towards high integration. Subsequently, research on development of operational types advanced powerfully, supported by national defense projects and space development projects in the U. S. In 1971, the microcomputer was developed, greatly reducing the size of computers and popularizing them widely.

The history of semiconductors in Japan began in 1951, when transistor technology was introduced from the U. S. At the time, virtually all semiconductor technology came into Japan in the form of technology introduction, and research and development work was concentrated on applying semiconductor technology

Fig. 1-17. Breakdown of IC Production in Japan and the U. S.

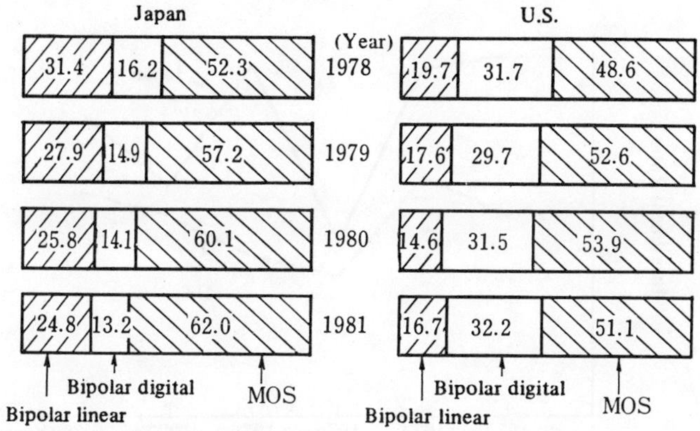

Note: Figures for 1981 are estimates
Source: The Electronics Industries Association of Japan

to consumer appliances, starting with the world's first mass production of transistor radios. Along with these efforts, noteworthy technological development work was carried out on tunnel diodes, static inductor transistors (SIT), and very large-scale integration (VLSI).

In the element technology necessary for the manufacture of IC's, Japan lags behind the U. S. in such fields as product planning and design, IC manufacturing equipment, and the technology of silicon wafers, but displays strength in production technology such as assembly, quality control, and inspection. In particular, Japan has surpassed the U. S. in automation of assembly work, a key factor behind its strong competitive power in IC's suitable for mass production. Also, as shown in Figure 1-18, Japan's level of investments in equipment and facilities is higher than in the U. S.

Although the semiconductor industry of Japan has depended upon the U. S. for many of its basic technologies, it has nonetheless improved productivity of IC's for consumer appliances, and this, together with excellent product quality control, has enabled it to win a good reputation internationally.

Among the many ways in which technological progress relates to production, in Japan this often takes the form of mechanization and automation of production and manufacturing processes intended to achieve lower costs and higher quality.

Of particular note in recent years is the appearance of high performance and multi-functional manufacturing machinery incorporating microprocessors, based on progress in semiconductor technology and generally referred to as 'mechatronics,' a word coined from mechanism and electronics.

Fig. 1-18. Investment Ratio of IC Equipment in Japan and the U. S.

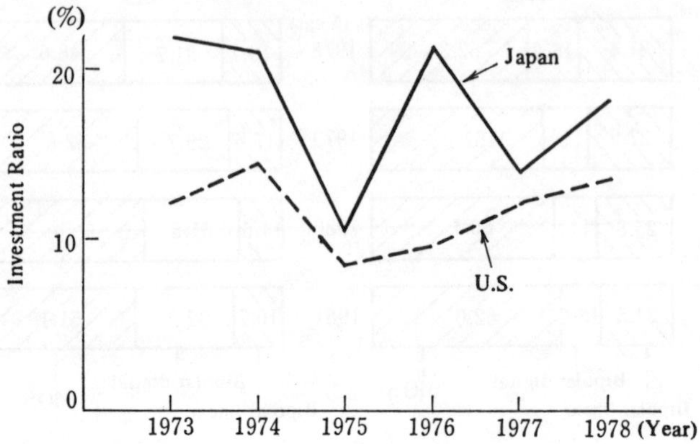

Note: Investment Ratio = Investment in Equipment/Sales
Source: The Economic Research Institute, Japan Society for the Promotion of Machine Industry

Figure 1-19 shows how electronics and mechatronics have grown and developed over the years, demonstrating that it has taken only a few years for new electronics technologies to be introduced into mechatronics, and that the trend is for the time lag to be reduced even further.

Industrial robots, considered the very quintessence of mechatronics, made their appearance in the 1960's, and along with progress in electronics technology their functions and performance have become increasingly sophisticated. Then, in the latter half of the 1970's, flexible manufacturing systems (FMS), in which various types of mechatronics equipment are group-controlled by computers, made their appearance.

At present, conversion to mechatronics is rapidly advancing in various industries in Japan, and has reached the point where it is being referred to as one of the factors behind the powerful international competitive power of Japanese industries. As shown in Figure 1-20, the monetary values of the production of NC (Numerical Control) machine tools and industrial robots have grown very rapidly in recent years. As Table 1-21 shows, already close to 70 percent of the major enterprises and nearly 19 percent of smaller enterprises have introduced them into their operations.

In this way, mechatronics, as exemplified by industrial robots, is rapidly becoming widespread, contributing greatly to improvements in productivity, product precision, and manpower savings, leading to a new era in which one production line can be used to produce a large variety of products in small quantities.

Fig. 1-19. Development of Mechatronics

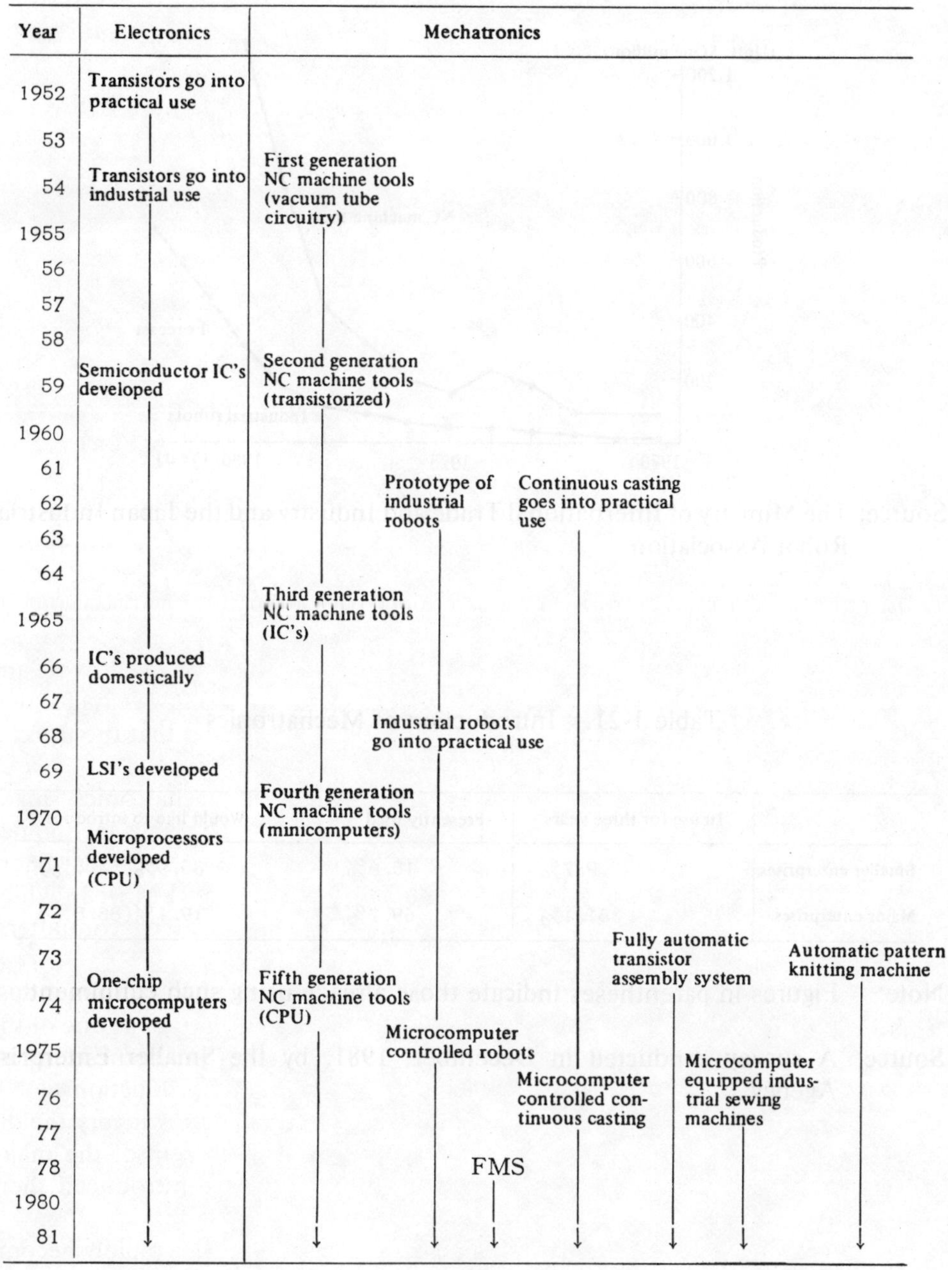

Source: The Economic Research Institute, Japan Society for the Promotion of Machine Industry

Fig. 1-20. Changes in Production Value of NC Machine Tools and Industrial Robots

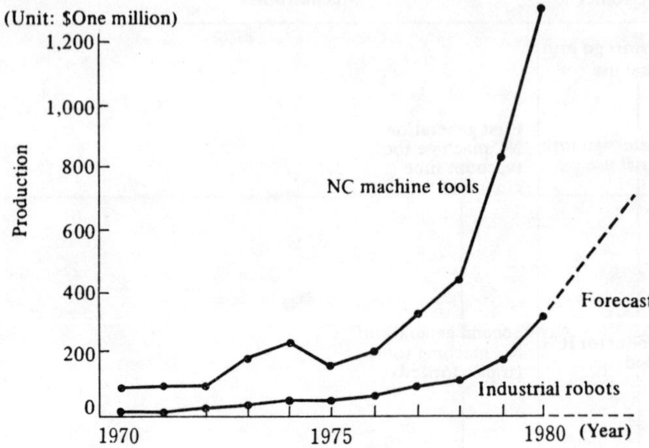

Source: The Ministry of International Trade and Industry and the Japan Industrial Robot Association

Table 1-21. Introduction of Mechatronics

	In use for three years	Presently used	Would like to introduce
Smaller enterprises	9.7%	18.6%	35.9%(54.5%)
Major enterprises	51.1%	69.7%	19.1%(88.8%)

Note: Figures in parentheses indicate those already using such equipment

Source: A survey conducted in December, 1981, by the Smaller Enterprise Agency

2. DEPENDENCY ON FOREIGN TECHNOLOGY

The degree of Japan's dependence on foreign technology can be seen statistically by looking at the number of technologies introduced from abroad and the amount of money paid for them. At the same time, comparable figures for exports show the amount of original Japanese technology contributed to other countries.

The introduction of technologies, which served to raise Japanese standards of science and technology after the end of World War II and which played a major role in the country's economic development, began in 1955 under a government approval system. Subsequently, reflecting the rapid development of Japan's economy toward one based on international trade expansion, with a relaxation of regulations on imporation the number of such introductions has grown sharply.

Figure 1-22 shows how this has progressed over the years. While the number of technologies imported increased in proportion to the growth in the economy, in recent years, the trend is a decreasing one. Classified by field, up to the latter half of the 1950's, the percentage in the chemical field was quite large, but since the first half of the 1960's, reflecting the rapid growth of processing-and-assembly type industries, the percentage in the fields of machinery and electrical machinery has become large.

In recent years, the standards of science and technology in Japan have approached those of the advanced nations of Europe and the U.S., and as a result Japanese technology can now be exported. As shown in Figure 1-23, while Japan's technological trade balance is still in the red, with growth in the value of exported technologies the balance has been slowly improving.

The technological trade balances of other leading nations show the U. S. in the black in a big way, while England and France are also in the black. However, although West Germany has consistently been in the red and continues to stay this way with little fluctuation, its deficit is smaller than Japan's.

Since most contracts covering technological trade require payment of royalties on sales of products incorporating the transfered technology, as sales increase this utilization fee also increases, so that in many cases contracts concluded in the past effect the technological trade balance of later years. Therefore, to grasp trends in technological trade at any given point in time, it is necessary to observe the balance of newly contracted items of technology. In terms of new contracts, Japan has been in the black since 1972.

Seen this way, although still in the red, Japan's technological trade balance is improving, while in the area of new contracts the balance has continued in the black, so that in the near future the country will be able to advance from an importer to an exporter of technologies.

But this trend does not mean that Japan is now on a par with other leading nations in terms of independently developed technologies. As Figure 1-24 shows, for other leading nations, the majority of technology exports are to developed nations, but in Japan most technology exports are to developing nations and

Fig. 1-22. Changes in Number of Technological Imports

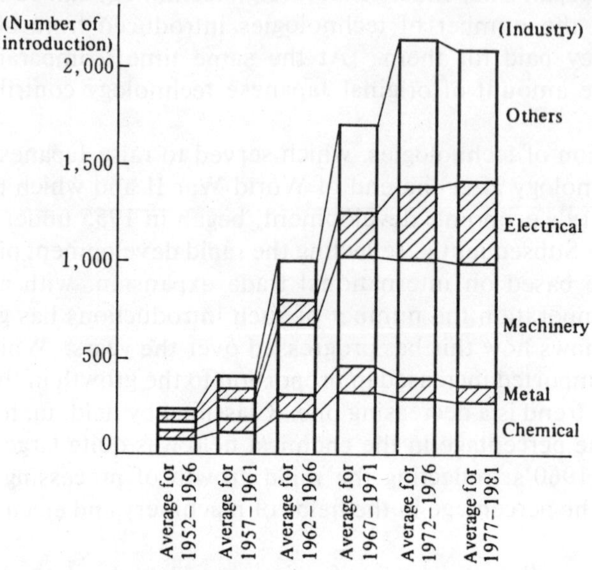

Source: The Science and Technology Agency

newly developed nations, and the proportion of technology exports to the advanced nations, approximately 40 percent, is still relatively low.

In other words, while Japan has often introduced technologies from advanced nations, adapted and improved them in original ways, and then exported these technologies to developing and newly developed nations, with few technologies which can be sold to advanced nations the country still lags in production of original, independently developed technologies.

In technological trade, the percentage taken by manufacturing industries is extremely high, from which it is possible to grasp a picture of overall trends. Changes in the value of technological trade in manufacturing industries are shown in Figure 1-25; changes in the proportion of technological trading taken by each industry are shown in Table 1-26; and changes in the technological trading balance by industry are shown in Table 1-27.

Fig. 1-23. Changes in Technological Trade Balances

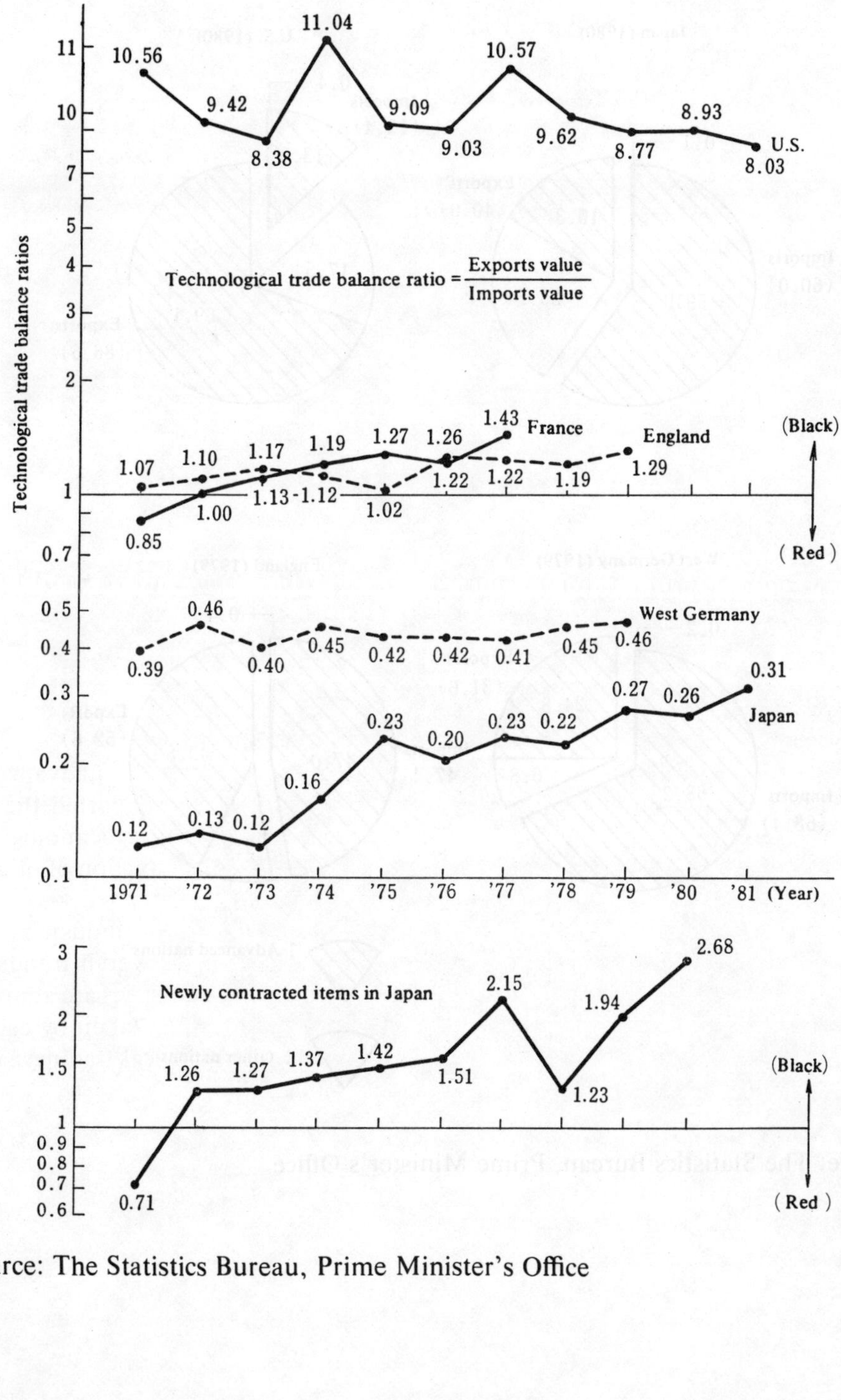

Source: The Statistics Bureau, Prime Minister's Office

Fig. 1-24. Technological Trade Balance Structures of Leading Nations

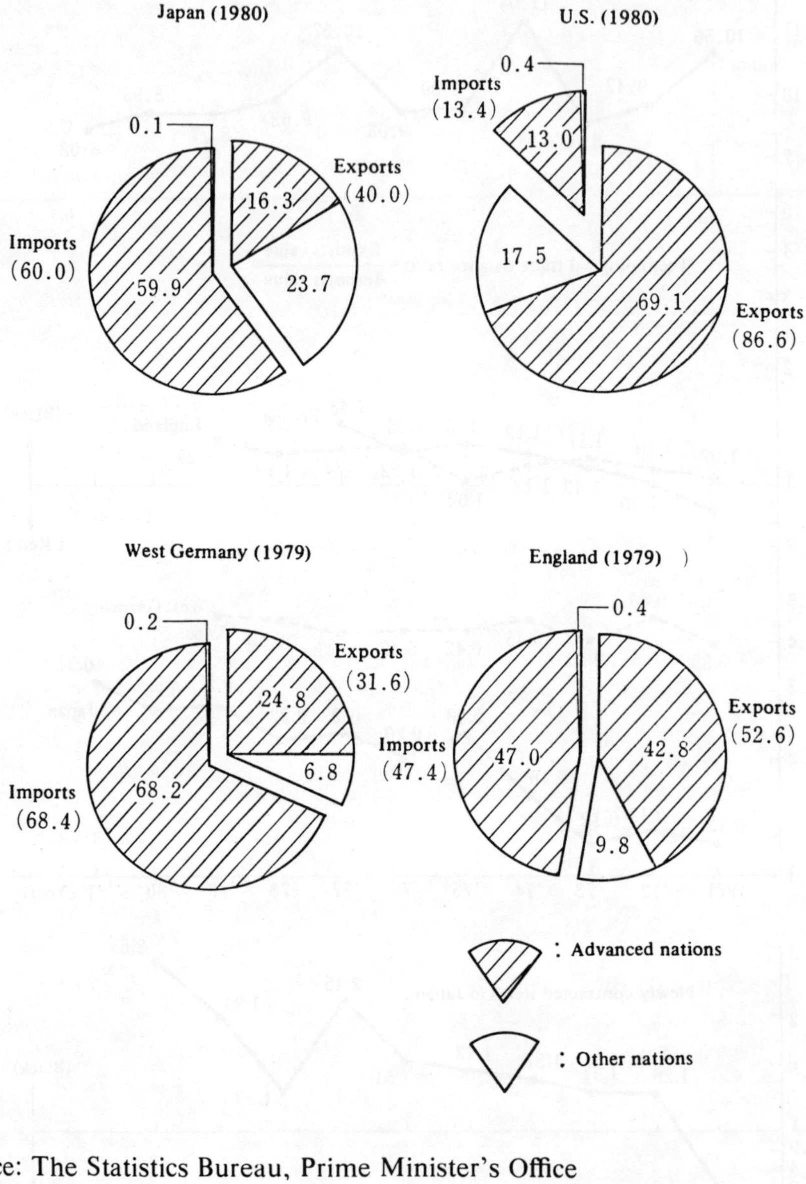

Source: The Statistics Bureau, Prime Minister's Office

Fig. 1-25. Changes in Values of Technological Trade in Manufacturing Industries

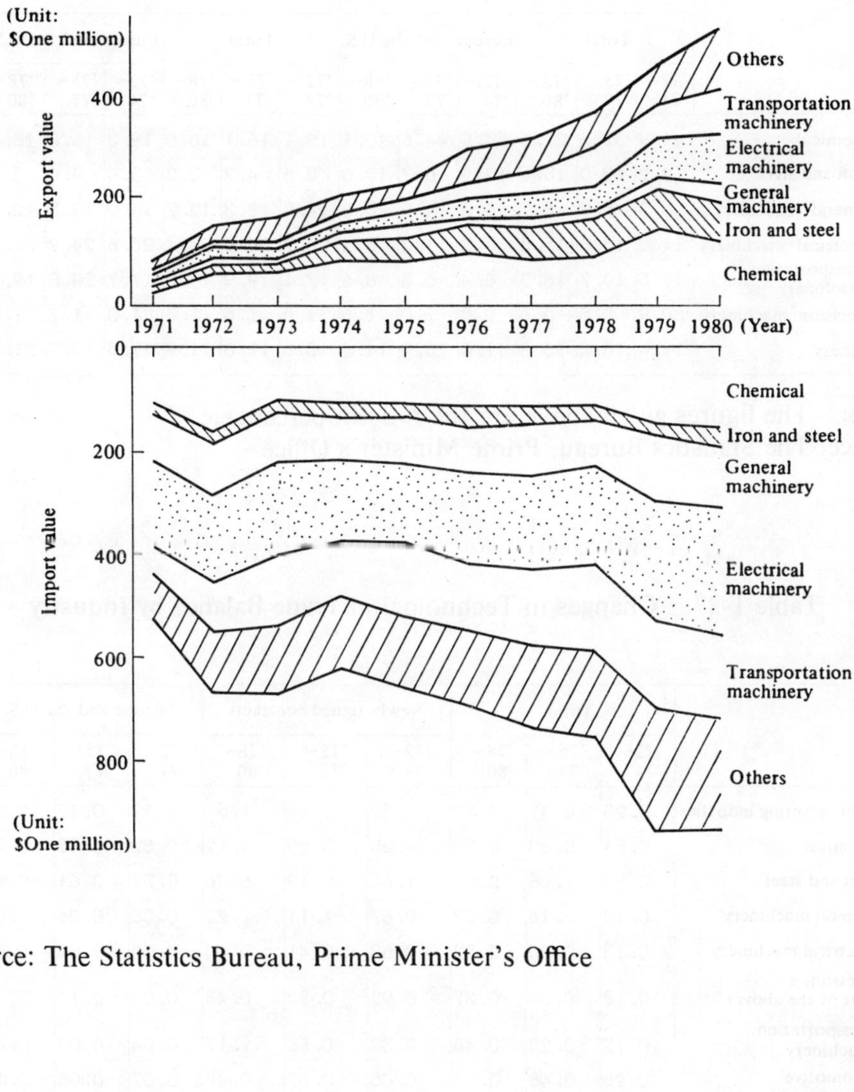

Source: The Statistics Bureau, Prime Minister's Office

Table 1-26. Changes in Proportion of Technological Trading by Industry

	Exports						Imports					
	Total			Europe and the U.S.			Total			Europe and the U.S.		
	'72~'74	'75~'77	'78~'80	'72~'74	'75~'77	'78~'80	'72~'74	'75~'77	'78~'80	'72~'74	'75~'77	'78~'80
Chemical	41.9	32.5	27.6	54.9	44.6	40.1	19.1	16.1	16.0	19.3	16.0	16.1
Iron and steel	12.2	19.0	15.5	12.9	16.7	13.6	3.6	4.7	3.2	3.6	4.6	3.2
General machinery	6.3	5.2	5.7	6.8	4.9	5.5	14.5	13.3	12.7	14.0	13.3	12.8
Electrical machinery	11.8	13.6	15.8	7.2	11.2	16.2	25.7	24.6	25.4	26.6	24.7	25.6
Transportation machinery	7.5	10.7	16.3	6.4	6.5	8.6	17.1	19.9	19.0	17.7	20.0	19.1
Precision machinery	0.9	0.5	0.8	0.8	1.0	1.6	1.5	1.6	1.8	1.3	1.7	1.8
Others	19.4	18.5	18.3	11.0	15.1	14.4	18.5	19.8	21.9	17.5	19.7	21.4

Note: The figures given above are the ratio in percentage
Source: The Statistics Bureau, Prime Minister's Office

Table 1-27. Changes in Technological Trade Balance by Industry

	Total			Newly signed contracts			Europe and the U.S.		
	'72~'74	'75~'77	'78~'80	'72~'74	'75~'77	'78~'80	'72~'74	'75~'77	'78~'80
Manufacturing industies	0.28	0.41	0.52	1.32	1.44	1.54	0.11	0.17	0.21
Chemical	0.61	0.83	0.90	4.25	3.29	4.17	0.30	0.48	0.51
Iron and steel	0.96	1.66	2.50	1.65	2.19	6.46	0.37	0.61	0.87
General machinery	0.12	0.16	0.23	0.67	1.11	1.72	0.05	0.06	0.09
Electrical machinery	0.13	0.23	0.33	0.63	0.41	0.50	0.03	0.08	0.13
Electronics (Part of the above)	0.18	0.31	0.37	0.92	0.83	0.48	0.04	0.11	0.18
Transportation machinery	0.12	0.22	0.45	0.32	0.66	1.17	0.04	0.06	0.09
Automotive (Part of the above)	0.28	0.48	0.82	0.08	0.57	0.45	0.07	0.06	0.11
Precision machinery	0.16	0.13	0.24	0.97	0.53	0.37	0.06	0.10	0.18

Source: The Statistics Bureau, Prime Minister's Office

During the period from 1971 to 1980, in the manufacturing industry the value of technological exports increased six-fold, and that of imports, 1.8-fold. The share of exports taken by the chemical industry has consistently been the highest, while in recent years the growing shares taken by the iron and steel industry, the electrical machinery industry, and the transportation machinery industry, approximately equal, immediately follow. The share taken by the electrical machinery industry in exports to Europe and the U. S. is now larger than the iron and steel industry. Meanwhile, virtually no changes can be seen in imports for each of the industries; the share taken by the electrical machinery industry is highest, followed by the transportation machinery industry, the chemical industry, and the general machinery industry, with the share of the iron and steel industry being quite low.

The chemical industry's share in the total value of technological exports, though gradually decreasing, is still quite high, and the amount to Europe and America accounts for more than 40 percent of the total. The overall value of technological trade is coming close to being balanced, and in particular the share of new contracts is well in the black. This is because of the good informational reputation of Japan's technology in fields such as urea manufacturing, synthetic resin production, and butane-butene separation. However, its technological trade balance with Europe and the U. S. continues to show a considerable deficit.

In the iron and steel industry, reflecting its high technological standards, the technological trade balance is in the black in a big way both for the overall total and new contracts. With the increase in exports of replacement equipment and facilities, blast furnace operation techniques, continuous casting technology, and patents and knowhow relating to high efficiency production systems, the technological trade balance vis-a-vis Europe and the U. S. is approaching a state of balance.

The international competitive power of the electrical machinery and transportation machinery industries has strengthened greatly in recent years, but, except for new contracts in the transportation machinery industry, the period from 1978 to 1980 still shows a deficit, with the deficits for Europe and America being particularly large. In other words, the strong international competitiveness of both industries can be mainly attributed to the high standards of production technologies, and not necessarily to advanced, independently developed technology.

The electrical machinery industry shows a great number of imports related to computer software, an area in which Japan has been lagging, while in the tansportation machinery imports of aero-space technology stand out. At the same time, however, the trade balances for both these industries are steadily improving along with advancements in electronics and automotive technology, and the automotive industry in particular approached a state of balance during the period from 1978 to 1980, and is rapidly improving.

As described above, while certain industries have achieved a technology trade balance, overall, because other industries are still importing a considerable amount of technology from abroad, the manufacturing industry as a whole is in the red. Although the situation is gradually improving, Japan is still oriented toward technological imports, and has not yet produced more than a relatively small number of independently developed technologies of worldwide interest.

3. LEVELS OF SCIENCE AND TECHNOLOGY

Although its level of science and technology has been rising in recent years, Japan's ability to create original technologies is weaker than that of Europe and the United States. In order to review Japanese standards of science and technology, the sources of Japanese technology, as well as the state of patent applications and registrations, publication of research papers, and international awards will be discussed.

(1) Origins of Japanese Technology

Table 1-28 shows the origin of selected industrialized technologies, based on a questionnaire sent to private enterprises in 1980. The technology taken up here all played important roles in the development of Japanese industries. It can be seen that the majority of these technologies are based upon overseas research and development work. It can also be seen that Japan has introduced technologies within several years of their industrialization abroad. There are also items, such as the MOS type IC, which Japan has succeeded in producing within a very short period after original development abroad. It is this sort of capability that has brought about the great improvement seen in the international competitive power of Japan's industries

Research and development in basic sciences and major technological innovations is extremely important in directing and stimulating progress in many other related fields. Table 1-29 lists a number of basic technologies, showing the year and place of origin.

From this table, it can be seen that the U. S. leads the world in both basic, theoretical principles and their experimental corroboration. Japan, as shown previously in Table 1-28, has also made certain contributions. In particular, optical fibers, theoretically established by Japanese researchers, are expected in the near future to be used in a wide range of applications. While there are areas in which Japan has made contributions to the world's science and technology, at least in recent years, from the overall standpoint, further efforts are still necessary.

Table 1-28. Origins of Selected Technology

(Industry) Technology	Development of original technology		Industrialization			
			Overseas		Japan	
	Year	Country	Year	(Country) Note	Year	Note
o (Food) Glutamin acid manufacturing process employing direct fermentation process	1956	Japan	–		1958	Industrialized
o (Synthetic fiber) Polymerization of ethylene at low temperature and normal pressure using catalyst	1953	West Germany	Around 1955	(West Germany) Industrialized	1959	Industrialized through technology introduced from West Germany
o (Synthetic fiber) Polypropylene polymerization technology employing catalyst	1954	Italy	1957	(Italy) Industrialized	1960	Industrialized using technology introduced from Italy
o (Pharmaceutical) Manufacturing of penicillin	1929	England (discovery of therapeutical effects)	During 1940's	(U.S.) Establishing of mass production technology	1944 1946	Extracted penicillin by original technology Industrialized using technology introduced from the U.S.
o (Petroleum) Modify low octane naptha into high octane gasoline using platinum catalyst	1947	U.S.	1947	(U.S.) Industrialized	1953	Industrialized using technology introduced from the U.S.
o (Petroleum) Produce high octane petroleum using fine dust fluid catalyst	1941	U.S.	1942	(U.S.) Industrialized	1954	Industrialized using technology introduced from U.S.
o (Iron and steel) Pure oxygen blowing converter	1952	Austria	1953	(Austria) Industrialized	1957	Industrialized by technology introduced from Austria
o (Iron and steel) Continuous casting technology	1935	Germany (for aluminum)	1940's	(European countries) Industrialized for steel	1955	Industrialized for steel by technology introduced from Switzerland
o (Electrical machinery) Video tape recorders for consumer use	1950	U.S. (for professional use)	1956	(U.S.) Production for professional use	1964	Production for consumer use
o (Electronic's) MOS IC's	1964	U.S.	1965	(U.S.) Industrialized	1965	Prototype of of MOS IC's
o (Automotive) Rotary engine	1954	West Germany	1964	(West Germany) Placed in practical use	1967	Placed in practical use by technology introduced from West Germany

Source: A survey of private enterprises conducted in January, 1980, by the Science and Technology Agency

Table 1-29. Development of Selected Basic Technologies

Technology	Development		
	Year	Country	Outline
Life Science			
DNA recombinant technology	1944	U.S.	Elucidated that DNA governs heridity
	1973	U.S.	Succeeded in transfering DNA of one cell to another
Technology Under Extreme conditions			
Superconductive technology	1911	Netherlands	Superconductive phenomenon discovered
	1957	U.S.	Superconductive theory established
New materials			
Optical fibers	1964	Japan	Graded index optical fiber theory
	1966	England	Loss reduction of optical fiber
	1970	U.S.	Low loss optical fiber developed
Electronics and communications			
Laser	1951	U.S.	Conception of microwave amplifier employing stimulated emission announced
	1960	U.S.	Solid state laser developed
Josephson junction	1962	England	Josephson effect predicted in terms of quantum theory
	1963	U.S.	Josephson effect verified through experiments
Tunnel diodes	1932	England	Existence of tunnel effect in semiconductors predicted
	1957	Japan	Tunnel diode invented
Transistors	1948	U.S.	pn junction transistor theory established
	1951	U.S.	pn junction transistor developed

Source: The Science and Technology Agency

(2) Patents

While differences in national patent systems distort simple numerical comparisons, the number of patents is nonetheless frequently used to appraise the results of research and development. Although the total number of patents registered by nationals of Japan, the U. S., West Germany, France, and England, as shown in Figure 1-30, has been decreasing in each of the five nations, the proportion held by Japan is on the increase. Since patent applications made to foreign nations are for items more strictly selected by the applicants, they more directly reflect the technological standards of each nation. As shown in Figure 1-31, in spite of the previously mentioned overall down-turn, again the proportion of applications from Japan is increasing, thus indicating a rise in Japanese technological standards.

Fig. 1-30. Changes in Patent Applications in Major Countries

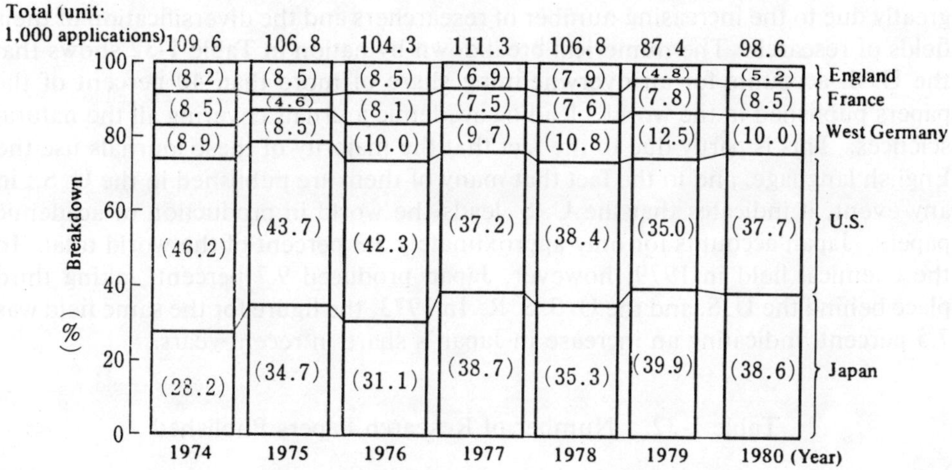

Note: Figures in parentheses indicate ratios; the width of each bar in the graph is proportionate to the total number for each year

Source: The Patent Agency

Fig. 1-31. Changes in Number of Patents Applications to Foreign Countries

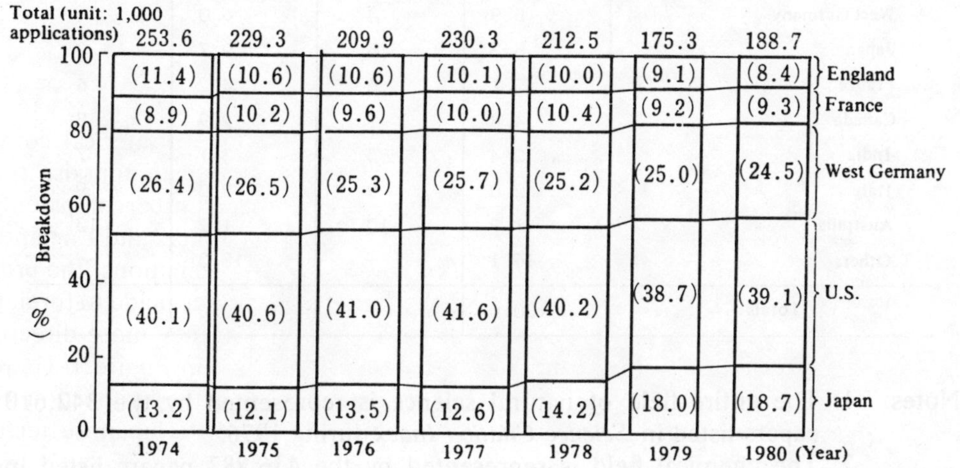

Source: The Patent Agency

(3) Research Papers

In recent years, the volume of research papers published in the world has risen greatly due to the increasing number of researchers and the diversification of their fields of research. The numerical breakdown by nation in Table 1-32 shows that the U. S. accounts for an overwhelming share of more than 40 percent of the papers published in the world's leading academic journals covering all the natural sciences. This is partly due to the fact that the majority of these journals use the English language, and to the fact that many of them are published in the U. S.; in any event, it indicates that the U. S. leads the world in production of academic papers. Japan accounts for only approximately 5.3 percent of the world total. In the chemical field in 1979, however, Japan produced 9.7 percent, taking third place behind the U. S. and the U. S. S. R. In 1973, the figure for the same field was 7.5 percent, indicating an increase in Japan's share in recent years.

Table 1-32. Number of Research Papers Published

Classification / Year / Item / Country	Entire field of natural sciences (SCI)[1] 1976 Proportion (%)	Entire field of natural sciences (SCI)[1] 1976 Ranking	Chemical field (CAS)[2] 1979 Proportion (%)	Chemical field (CAS)[2] 1979 Ranking
U.S.	41.9	1	27.5	1
England	8.2	2	5.9	5
U.S.S.R.	6.9	3	19.0	2
West Germany	5.9	4	6.0	4
Japan	5.3	5	9.7	3
France	5.1	6	4.2	6
Canada	4.4	7	2.9	8
India	2.4	8	3.1	7
Italy	2.0	9	2.2	9
Australia	1.9	10	1.2	13
Others	16.1	—	18.3	—
Totals	100.0	—	100.0	—

Notes: 1. The entire field of natural science is represented by the 342,610 papers listed in *Science Citation Index* during 1976.

2. The chemical field is represented by the 436,887 papers listed in *Chemical Abstracts* during 1979.

3. Country is that of the organization to which the first author belongs.

Source: *Library and Information Science*, No. 16, 1979, and *CAS Today*

From a qualitative standpoint, publications more often cited by other researchers can be considered of higher quality. According to an analysis conducted by the Institute for Scientific Information of the U. S., of the 1,000 scientists worldwide whose publications had been most cited, Japanese scientists account for only 19, or 1.9 percent. Moreover, approximately half of the researchers belong to American institutes. While it is true that most research in Japan is published in the Japanese language, making translation an additional, difficult task, based on this analysis of high level research work Japan's contribution to the international community of scientists is still low.

(4) International Science Awards

The number of scientists who receive major awards for their achievements reflects to a considerable extent the scientific and technological standards of a nation. For ease of comparison, among numerous awards the number of recipients of only the Nobel Prize in physics, chemistry, medicine, and biology and of the Fields Award in mathematics are shown by country in Table 1-33. According to this analysis, since establishment of these awards the U. S. has produced the overwhelmingly largest number of recipients, followed by England, Germany, and France. Japan, with four recipients of the Nobel Prize and two of the Fields Award, is nowhere near the other leading nations, but is instead comparable to nations such as Italy and Belgium.

In summary, while it is true that the level of basic science and technology in Japan is gradually improving, there are still many areas in which Japan lags behind the advanced nations of Europe and, in particular, the United States.

4. FEATURES AND BACKGROUND

In contrast to the previous sections, this section will look at the features and background of Japanese technology as seen from the results of questionnaires sent to private enterprises.

Figure 1-34 summarizes opinions of specialists in private enterprises on reasons for Japan having attained its current technological standards, divided into technology relating to product quality and technology relating to production processes. It will be seen from this that the degree of Japan's dependence on foreign technology is quite large.

Table 1-33. Number of Recipients of International Science Awards

Award / Country	Nobel Prize 1901～1982	Fields Award 1936～1978
U.S.	126	8
England	63	3
Germany (both East and West Germany included)	49	0
France	22	4
Sweden	15	1
U.S.S.R.	10	2
Netherlands	9	0
Switzerland	9	0
Austria	8	0
Denmark	7	0
Italy	5	1
Belgium	5	1
Japan	4	2
Others	23	2
Total	355	24

Source: The Science and Technology Agency

Fig. 1-34. How Japan Attained Its Current Technological Standards

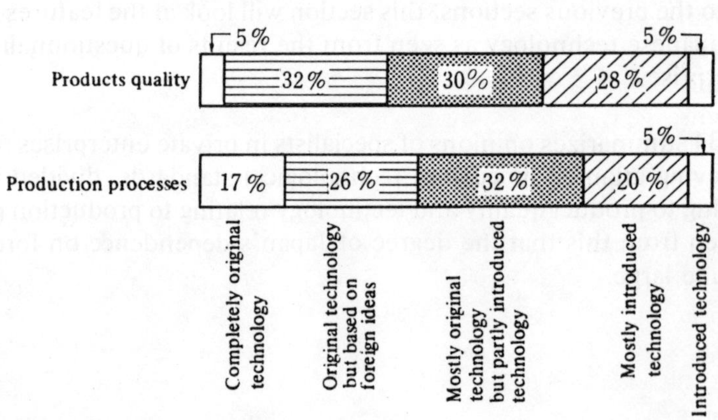

Note: Ratio of respondents

Source: Questionnaires commissioned by the Agency of Industrial Science and Technology, Ministry of International Trade and Industry, in December, 1978

Next, breaking down factors behind technological development in Japanese industries during the latter half of the 1970's into innovations in production processes, product performance, and management techniques, Figure 1-35 shows that overall much credit is due to innovations in production processes, but that the proportion of innovations in product performance is small. This no doubt relates to the fact that, as shown in Figure 1-34, the degree of dependence upon foreign countries is larger in the field of product technologies than in production technologies.

Fig. 1-35. Factors in Technological Development

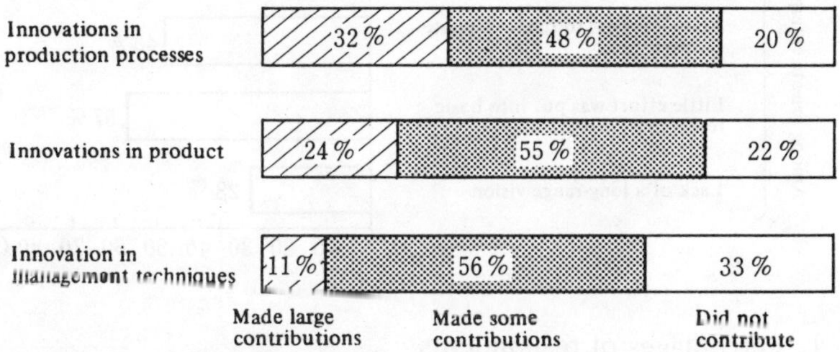

Note: Ratio of respondents
Source: Questionnaires conducted in February, 1982, by the Economic Research Institute, Japan Society for the Promotion of Machine Industry

In a 1980 survey of management executives responsible for R&D in private enterprises, approximately three quarters of all respondents gave the opinion that Japan is behind Europe and the United States in its level of original, innovative research. Figure 1-36 presents reasons given by these executives for this situation, and shows that the largest percentage referred to the existence of already developed technologies abroad, which could be imported and then improved upon, as a reason for not carrying out creative research domestically.

The fact that little effort was put into basic research was pointed out as a major reason by the enterprises themselves. It is quite clear that if new products are created one after another through technological innovations based on new principles drawn from creative research, market superiority can be maintained, but to bring about such results it is necessary to build up over a long period a wide range of time-consuming but steadily accumulating basic research. But in Japan, only short-term results have been pursued, and for the above reasons priority has not been placed on basic research. This situation also relates to replies to the survey which stated that there was no willingness to fairly evaluate creativity and that a long range vision was lacking.

Fig. 1-36. Reasons Why Japan Lags in Innovative and Creative Research

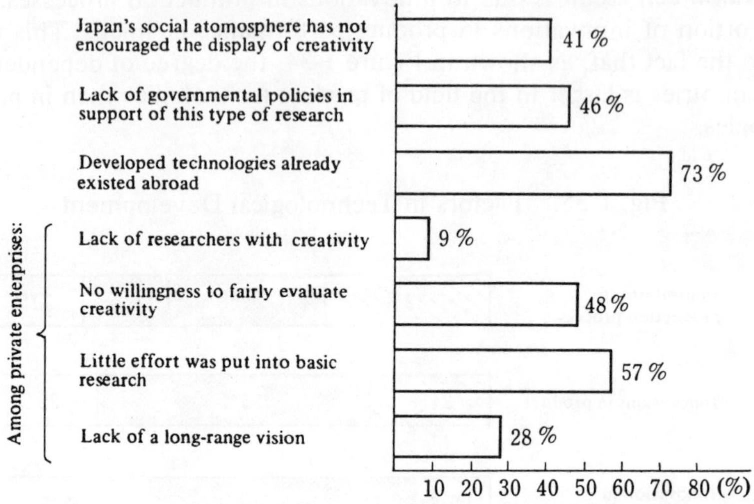

Notes: 1. Percentages of respondents
2. Multiple answers

Source: Questionnaires commissioned by the Science and Technology Agency in July, 1980

In addition, the fact that Japan's social atmosphere has not encouraged the display of creativity was pointed out. Probably because so much priority has been placed on short-term results, the adaptation of fully-grown technologies was stressed more than the nurture of newly-sprouted processes, thus giving less importance to the unique ideas of creative individuals and more to the power of well-organized groups to attain pre-determined social objectives. The final result was a tendency to treat creativity and individuality lightly, neglecting the more long-range fruits of original basic research.

An additional major reason mentioned was the lack of government policies in support of this type of research. Although the government has supported basic research at universities or national laboratories, the results suggest that the private sector is seeking government policies to support creative research and development at private enterprises as well.

As will be discussed in the next chapter, with demands for the development of original technology increasing, changes in both industry and society necessary to meet these demands are expected to take place.

CHAPTER 2

DEVELOPMENT OF ORIGINAL TECHNOLOGY

1. RISING DEMAND

Demands for development of original technology in Japan are rising, reflecting various social and political factors. In the background is the growing expectation that major technological breakthroughs will open up new possibilities of resolving the many problems facing the world, such as the stagnating economy and dwindling resources.

Under these circumstances, international calls for Japan to contribute more to scientific and technological progress are on the rise. In a 1982 survey of six EC nations conducted by the Ministry of Foreign Affairs, many respondents called on Japan to contribute to international society by assisting the development of world economy and technology. Similarly, the importance of science and technology in revitalizing the troubled world economy was stressed at the Eighth Summit Conference, held in France in June, 1982, in response to which Japan made a commitment to actively pursue international cooperation in research and development.

Furthermore, a more immediate problem is the increasing difficulty of importing foreign technology, on which Japan has greatly depended. As Japan approaches the level of Europe and the United States, the amount of new technology available has dropped, and foreign firms have grown more reluctant to make disadvantageous agreements with competing Japanese companies. Foreign companies now demand stricter terms for technology trade, such as cross licensing, capital participation, stiffer royalties, and restrictions on future export markets.

2. STEPS TOWARD ORIGINAL TECHNOLOGY

This section will overview trends in research expenditures and personnel, as well as research activities in the private and government sectors, from the standpoint of the development of creative technology. More detailed discussion of these matters will be taken up in Chapter 3.

(1) Research Expenditures and Personnel

In recent years, Japan has narrowed its technological gap with Europe and the United States, and in the process strengthened the international competitive power of its industries. One of the factors supporting this growth is the country's active investment in research and development. The size of Japan's research expenditures is now the third largest in the free world, following the U. S. and West Germany. The ratio of research expenditures to the national income, given in Figure 2-1, rose rapidly in 1979 and 1980, when it reached 2.42 percent, the highest ever in Japan. However, when comparing this ratio to those of other countries (in 1978, the only year for which full statistics are available), Japan is still behind West Germany, the U. S., and England.

These research and development investments can be broken down into those made by private industry or by the government. Compared to Europe and the United States, the share of private research expenditures is higher in Japan.

As shown in Figure 2-2, the year-to-year rate of growth from 1970 on was around 20 percent for both private and governmental research expenditures (except for 1971, when spending by the private sector was exceptionally low because of the recession). From 1975, however, following the onset of the oil crisis, the growth rates for both sectors dropped to around 10 percent, and although there were some annual fluctuations, generally speaking both sectors showed similar growth rates in research expenditures. But this trend started to change in the beginning of 1979, and research expenditures made by the private sector grew 15 percent in 1979 and 18 percent in 1980, while those made by the government dropped 11 percent and 9 percent, respectively.

Fig. 2-1. Ratio of Research Expenditures to National Income

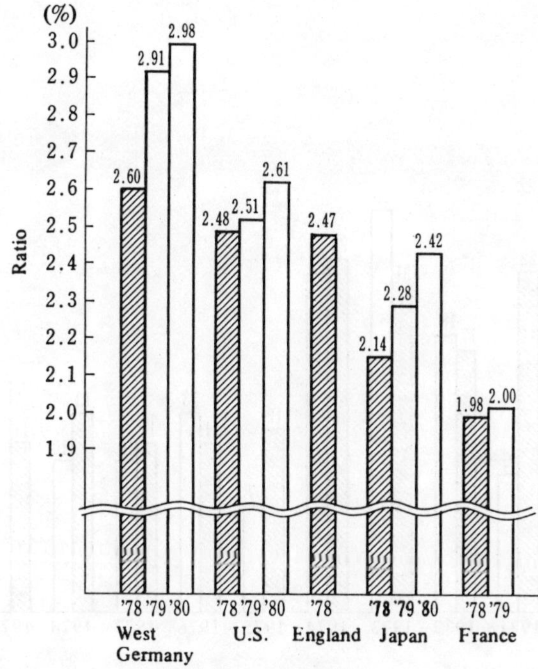

Note: Research expenditures for the U. S., England, and France also include those for cultural and social sciences.

Source: The National Science Foundation, the OECD, etc.

Fig. 2-2. Growth Rates of Research Expenditures in Government and Private Sectors

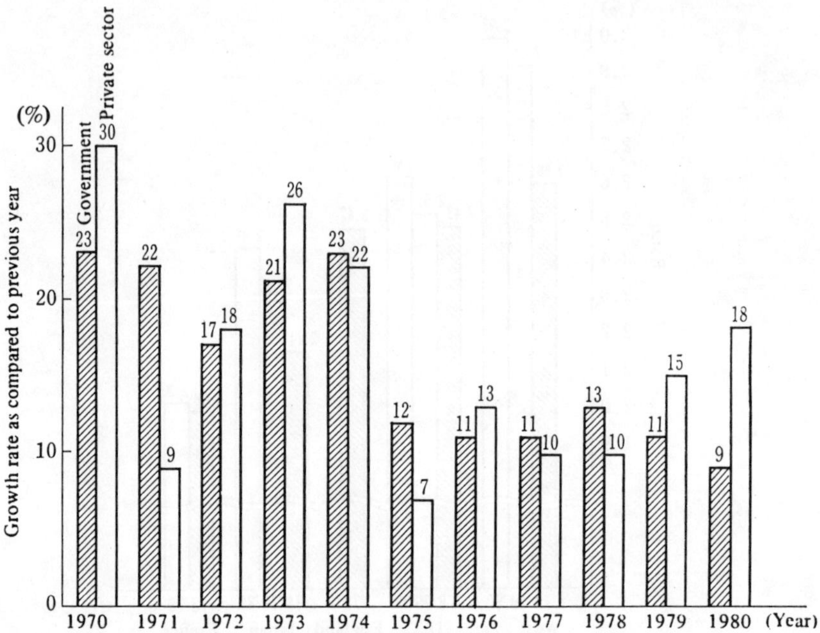

Source: The Statistics Bureau, Prime Minister's Office

Research and development work can be classified into basic, applied, and developmental research. It is the first of these, basic research, which is indispensable for the development of creative technology. But when changes in research expenditures are broken down into these three categories for the five-year period from 1976 onwards as in Figure 2-3, it can be seen that for 1979 and 1980 growth in basic research expenditures dropped as compared to growth in developmental and applied research expenditures. Although up to 1978 basic research expenditures grew relatively steadily, the growth rate in 1979 was negative and in 1980 was zero. In spite of increases in the private sector for both these years, this resulted from negative growth in basic research expenditures at universities and research laboratories, which account for about three quarters of total basic research expenditures in Japan.

In the free world, Japan is second only to the United States in the number of researchers, approximately 320,000 in April, 1981. Comparing changes in the number of researchers per 10,000 capita over the past 10 years in the U. S., West

Fig. 2-3. Growth Rates of Research Expenditures

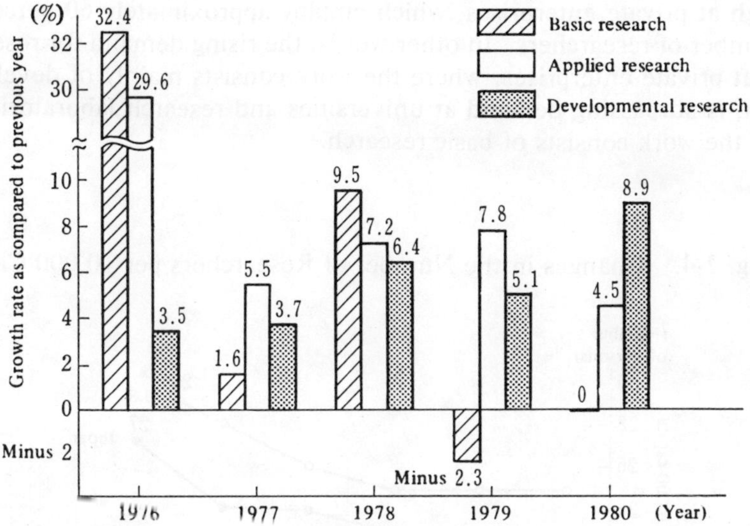

Source: The Statistics Bureau, Prime Minister's Office

Germany, and France, Figure 2-4 shows that although the number in each of these nations has tended to increase, the growth in the figure for Japan is particularly conspicuous, reaching 27 persons in 1981, nearly the same as the U. S. But research expenditures are small in relation to the large number of researchers; per researcher, the figure for Japan is at a low level, approximately 50 percent of that for West Germany and France, and approximately 70 percent of that for the U. S.

Classified by organization, approximately 60 percent of all researchers in Japan are affiliated with private enterprises, 10 percent with research laboratories, and 30 percent with universities. However, the increase in the number of researchers at private enterprises between 1980 and 1981 was higher than at universities and research laboratories, which have always played a major role in basic research.

In order to discover more about the background of these researchers, university graduates and post graduates in Japan in the area of natural sciences were classified according to their speciality. Figure 2-5 shows that the largest share, 61 percent, is taken by engineering majors, while majors in the physical science account for only 10 percent.

Although differences in national academic systems make direct comparisons very difficult, the situation in the U. S., England, and West Germany, where the number of physical science majors surpasses those in engineering, is in pronounced contrast with Japan.

The fact that the number majoring in engineering, an applied science, far surpasses the number majoring in physical science, a basic science, is closely related to demand for research personnel in Japan. Demand for engineering majors is very high at private enterprises, which employ approximately 60 percent of the total number of researchers. In other words, the rising demand for research personnel at private enterprises, where the work consists mainly of developmental research, is surpassing demand at universities and research laboratories, where most of the work consists of basic research.

Fig. 2-4. Changes in the Number of Researchers per 10,000 Capita

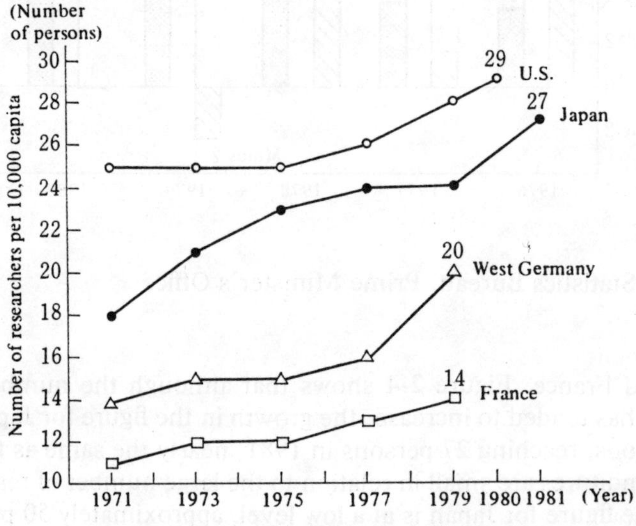

Source: The Statistics Bureau, Prime Minister's Office

Fig. 2-5. Breakdown by Speciality of Natural Science Graduates

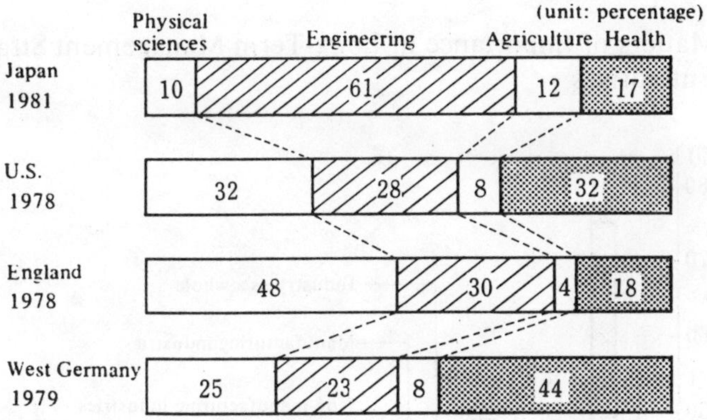

Source: The Ministry of Education

(2) Research Activities

Figure 2-6 shows the degree of importance private enterprises put on research and development. According to this survey, industry as a whole lists "improvement of ability to produce new technology" as the most important item from the standpoint of long-term management strategy. The proportion is particularly high in manufacturing industries.

Next, concerning prospective investments in research and development, Figure 2-7 shows that 15 percent of all industries expect that their 1985 ratio of research expenditures to sales amount will "increase considerably," and 51 percent report that the ratio will "increase somewhat." In other words, a total of two thirds of all industries surveyed recognize a need to continue to increase their investments in research and development. By industry, the proportion of enterprises that believe that the ratio will "increase considerably" is quite high in the electrical equipment, automotive, shipbuilding, and precision instruments industries, showing that they take a positive attitude towards research and development.

Presenting trends in demand for researchers, Figure 2-8 shows that approximately half of the enterprises would like to hire graduates who have majored in electronics and communications, where the most intense technological progress is expected. Also, approximately 10 percent of the enterprises want to hire graduates of doctoral courses. Here again, those in electronics and communications comprise the largest number, followed by the fields of pharmacology and biology.

In addition, demand for persons trained in chemistry, mathematics, and physics is also strong.

Fig. 2-6. Matters of Importance in Long-Term Management Strategy at Private Enterprises

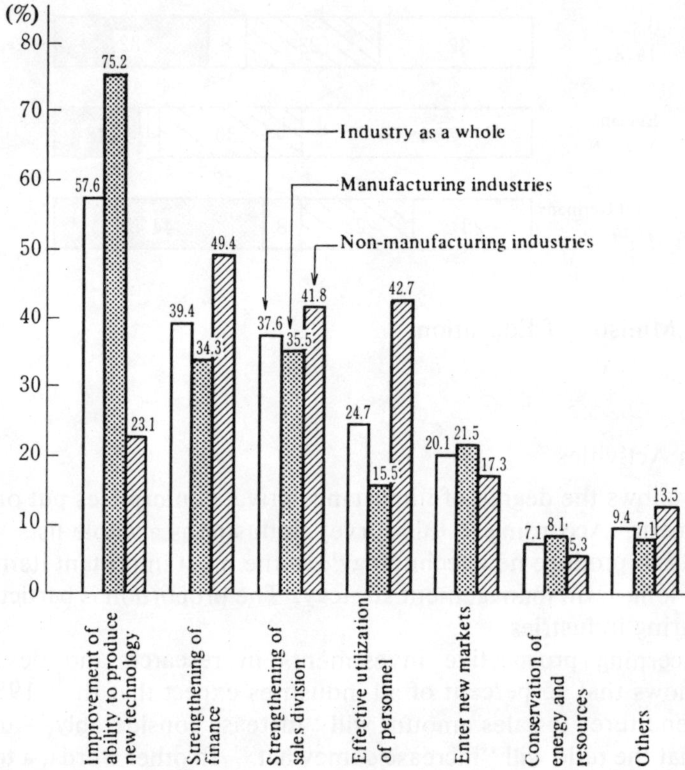

Note: Percentage of corporations that responded to questionnaire. Since multiple replies were received, total exceeds 100.

Source: Questionnaires conducted in 1981 by the Economic Planning Agency

Fig. 2-7. Projected Ratio of Research Expenditures to Sales (for 1985)

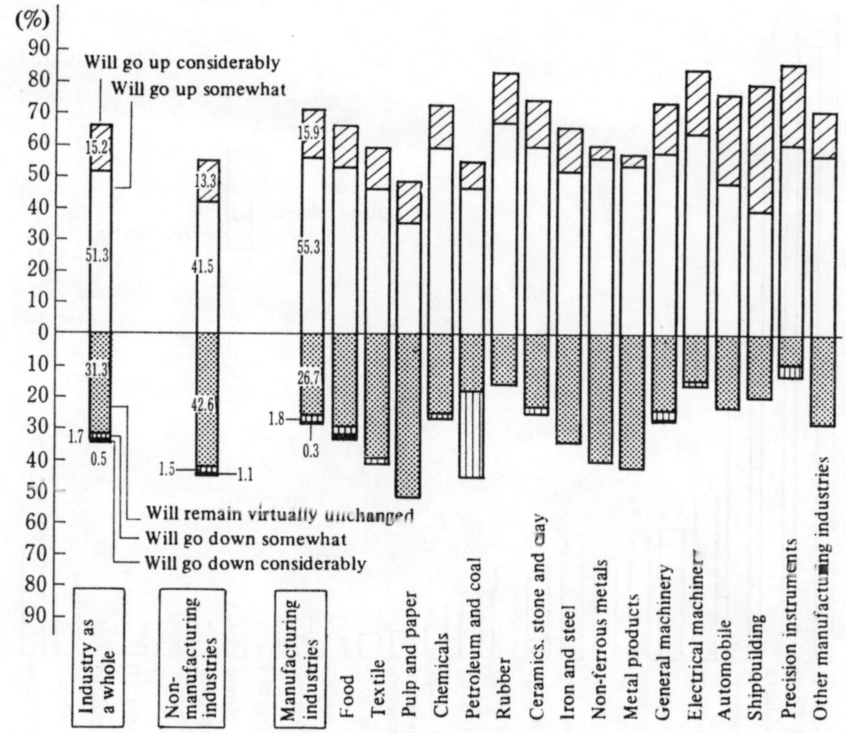

Note: Percentage of corporations that responded to questionnaires
Source: Questionnaires conducted in 1981 by the Economic Planning Agency

Fig. 2-8. Fields in which Private Enterprises Want to Increase Employment of Graduates

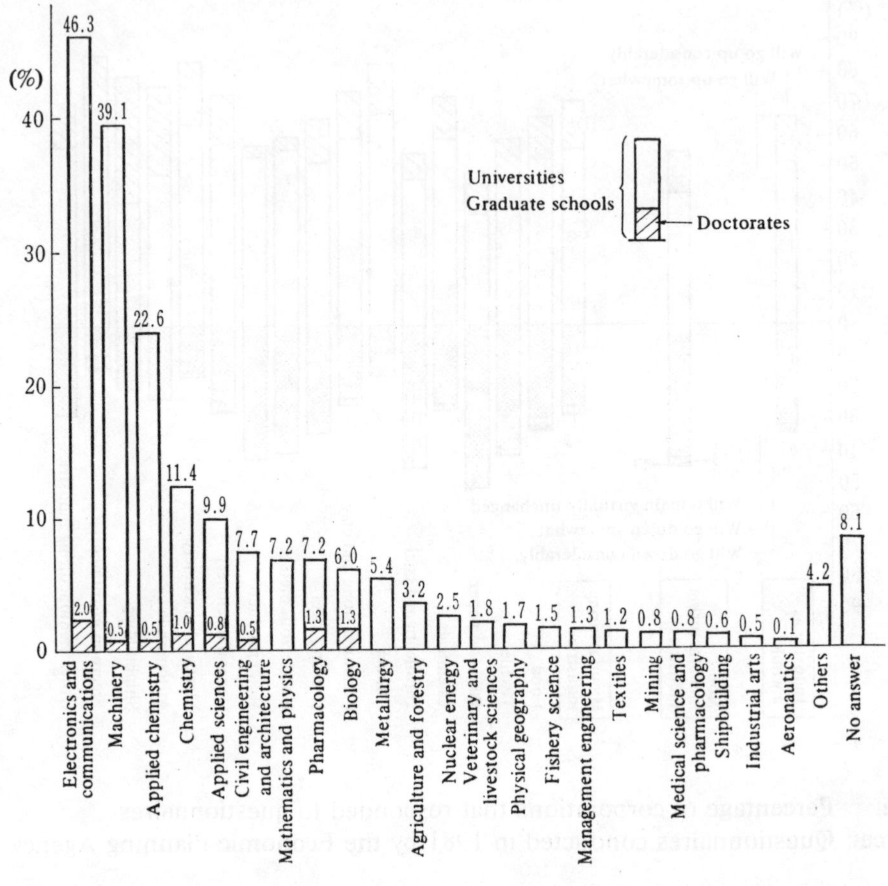

Note: Percentage of corporations that responded to questionnaire. Since multiple replies were received, totals exceed 100.
Source: A study conducted in 1981 by the Science and Technology Agency

According to a 1982 survey by the Economic Planning Agency, 68 percent of all industries (78 percent of the manufacturing industries) expect to emphasize independently developed technology in their future research and development work. As shown in Figure 2-9, reasons given — such as greater profitability, lack of useful technologies to import, and difficulties in reaching agreement — reflect the fact that overseas technologies which private enterprises have depended upon in the past can no longer be as easily obtained.

Fig. 2-9. Reasons for Placing Priority on Independently Developed Technology

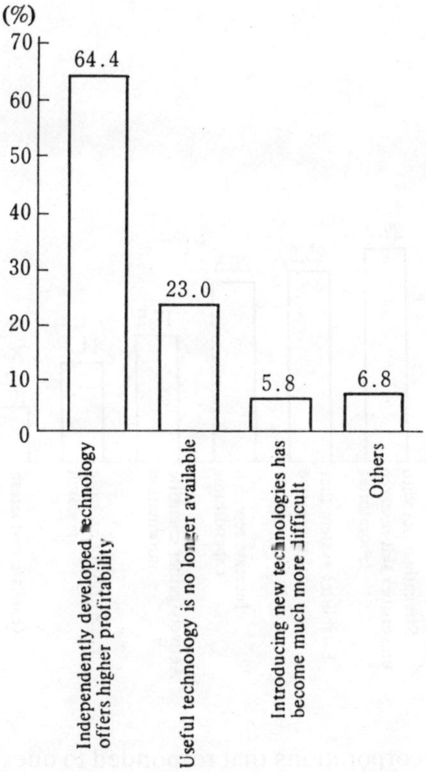

Note: Percentage of corporations that responded to questionnaire
Source: Questionnaires conducted by the Economic Planning Agency

Next, Figure 2-10 shows how private enterprises intend to promote development of independent technology. It indicates that in addition to obvious measures such as increasing personnel and research expenditures, various other steps such as establishing long-term plans and working in collaboration with other organizations will be necessary.

Of Japan's total research expenditures, the government (both national and local) invests slightly more than a quarter, of which approximately 80 percent is borne by the national government. The role of the government is to improve standards in the country's science and technology as a whole, and to assist the efforts of private enterprises.

The government promotes (1) basic research at universities; (2) creative and basic research with high risks; (3) large-scale projects such as development of

Fig. 2-10. Measures to Promote Development of Independent Technology

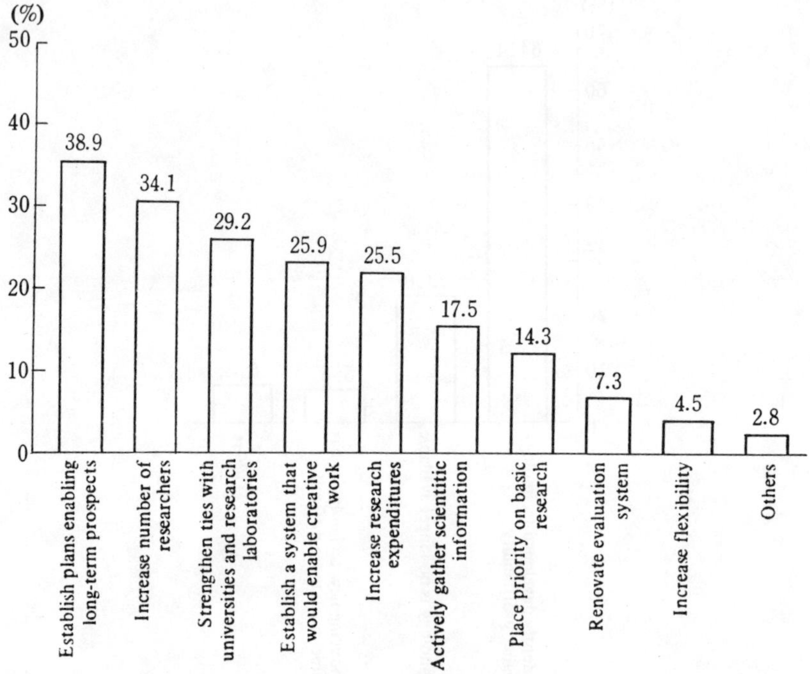

Note: Percentage of corporations that responded to questionnaire. Since multiple replies were received, total exceed 100.

Source: A Survey conducted in 1981 by the Science and Technology Agency

nuclear energy, space, and marine technology; (4) research on projects of social and economic importance requiring immediate solution; (5) research projects for the public such as disaster prevention, health and welfare; (6) research in fields such as agriculture where it is not possible for persons engaged in the work to conduct research themselves.

In 1981, to emphasize creative, independently developed technology, the government established three programs. The first, the System for Exploratory Research for Advanced Technology, is an innovative flexible research system where researchers from private enterprises, universities, and public research laboratories, including some from overseas, are gathered together under outstanding project leaders. The second program, the Research and Development Project of Basic Technology for Future Industries, coordinates parallel development work at private sectors, universities, and national laboratories. The third program, the Special Coordination Funds for Promotion of Science and Technology, supports the development of creative technology, projects requiring the cooperation of several laboratories, and coordination of private, academic, and government sectors. An outline of the first two programs is given in Table 2-11.

Table 2-11. Outline of the System for Exploratory Research for Advanced Technology and the Research and Development Project of Basic Technology for Future Industries

	The System for Exploratory Research for Advanced Technology	The Research and Development Project of Basic Technology for Future Industries
Objectives	The creation of new concepts and ideas that will serve as basis of future science and technology	Research and development of innovative industrial technology from "budding," or the theoretically proven stage, to "sapling," or the stage of proving their practicality
Projects	Ultra-fine Particle Amorphous & Laminar Materials Fine Polymers Perfect Crystals (Established in 1981) Bio-holonics (Established in 1981)	New Materials — 6 subjects Bio-technologies — 3 subjects New Electronic Devices — 3 subjects (Established in 1981)
Method of research	o Flexible research system in which researchers from private, national, as well as overseas laboratories are gathered o Foundation of research group within Research Development Corporation of Japan using facilities of private research laboratories	o Parallel development system through cooperation among private, academic, and government sectors o Commissions research to private enterprises
Research period	In principle, five years	Overall program of approximately 10 years, broken down into three stages
Budgets	1981 $2.4 million 1982 $8 million	1981 $10.8 million 1982 $19.2 million

3. PROGRESS IN VARIOUS FIELDS

For Japan to develop original technology in the future, it is important not only to direct efforts toward research and development of innovative technology but also to replenish and reinforce existing science and technology. The results will exert far-reaching effects in many fields and contribute to the improvement of the level of science and technology as a whole.

From this point of view, the present section will discuss progress in the following seven fields: (1) nuclear fusion; (2) space technology; (3) marine technology; (4) life science; (5) technology using extreme conditions; (6) new materials; and (7) electronic and communication technology. At the same time, targets of research and development in each field and progress achieved to date will be introduced, with comparisons made with the level of technology in other countries.

Nuclear Fusion

In the case of energy produced from nuclear fusion, the main ingredient in the fuel can be obtained from sea water, permitting an almost permanent supply of energy once the technology has been put into practical application. Therefore, nuclear energy is expected to become the ultimate energy source of the future. However, there are still a great many technological hurdles to be cleared, and much pioneering work remains to be done. Moreover, the large-scale equipment and huge investments required from the initial stage of research make it necessary to develop technology through international cooperation. Such being the case, very large-scale joint projects have already been started, in which Japan is participating. Meanwhile, the high level of scientific and technological accomplishments to be achieved in the course of research and development of nuclear fusion technology are expected to bring about secondary effects, accelerating new developments in other fields.

(1) Outline of Research and Development

Nuclear fusion results when nuclei of hydrogen, heavy hydrogen, helium and other light atomic elements are fused, converting to far larger atomic nuclei and simultaneously emitting energy. For instance, the amount of energy generated in nuclear fission by one gram of fuel consisting of the same number of heavy hydrogen and tritium atoms is roughly equivalent to the amount of energy obtained from combustion of about 8,000 liters of petroleum, or the amount of energy generated by nuclear fission of five grams of uranium.

The most familiar example of nuclear fusion is the sun. At the core of the sun, atomic nuclei of hydrogen and helium are constantly reacting and emitting energy of about 20 million Kelvin under densities as high as about 150 g/cm^3. In creating nuclear fusion on the earth, it is necessary to confine the fuel materials to a limited space. The higher the temperature and density, and the longer the confinement time, the larger is the amount of energy obtained. The theoretically required temperature for nuclear fusion extends from about one hundred million to several hundred million Kelvin. At such extraordinarily high temperatures, it becomes impossible for the materials, then in a plasma state, to be confined in metal or other conventional containers. Consequently, research is now being carried out on the use of a magnetic field to confine the materials. In order to realize the practical application of nuclear fusion as a future energy source, the output energy generated by the reaction must be larger than the input energy needed for the confinement and heating of plasma. The target of current research is to materialize nuclear fusion reactions between heavy hydrogen (D) and tritium (T), which require a comparatively low temperature and thus a comparatively small amount of input energy, as illustrated in Figure 2-12.

The technology required for a nuclear fusion reactor can be roughly classified into reactor core technology and reactor engineering technology. Reactor core technology refers to technology which permits stable confinement of fuel materials and causes a nuclear fusion reaction. Reactor engineering technology refers to technology which improves the durability and economics of materials and equipment, production and supply of fuel, and extraction of generated energy.

So far in Japan, research on nuclear fusion has been concentrated mainly on technology related to the reactor core. Between 1955 and 1965, various research activities were initiated at universities to study the physical phenomena of plasma. In 1968, Japan's Atomic Energy Commission, in order to promote research on nuclear fusion with the major role to be played by the Japan Atomic Energy Research Institute, initiated the "Basic Plan for Research and Development of Nuclear Fusion." Research and development conducted from 1969 through 1974 in accordance with this plan marked the first stage of nuclear fusion in Japan. The safe confinement of plasma of the order of several million Kelvin, a target for this stage, was accomplished using JFT-2 Tokamak equipment by the Japan Atomic Energy Research Institute, and was well received internationally. Since 1975, Japan has been in the second stage of this basic plan.

The purpose of this basic plan is mainly, (1) to develop a Tokamak-type critical plasma experimental device for the purpose of attaining critical plasma conditions (described later), a prerequisite for materializing nuclear fusion; and at the same time, (2) to carry out research and development of high beta plasma and a non-circular, cross-section plasma Tokamak device intended to improve the efficiency of the confinement of plasma; and further, (3) to carry out research and development pertaining to superconductive technology, heavy hydrogen processing technology, material engineering and other reactor engineering technologies required

Fig. 2-12. Combination of Fuel Materials and Degree of Potential of Nuclear Fusion Reaction

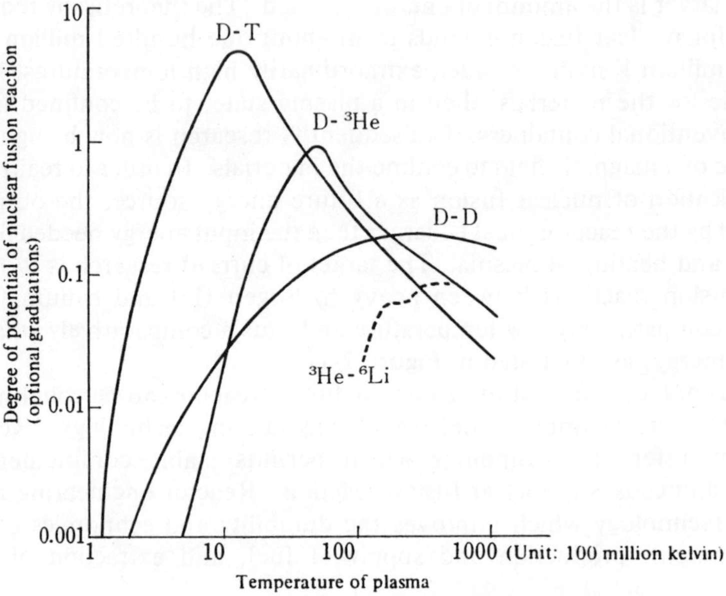

Note: D: Heavy hydrogen; T: Tritium, ^{3}He: Isotope of helium; ^{6}Li: Isotope of litium

Source: The Japan Atomic Energy Research Institute

for the development of nuclear fusion reactors; and finally, (4) to incorporate the results of the above into the third and subsequent stages of research and development activities. In compliance with this plan, work is now being done by the Japan Atomic Energy Research Institute and other national research laboratories. Moreover, basic research on methods of confinement is being carried out under this plan by universities and other institutes in an effort to discover new confinement techniques using mirrors, stellarators, heliotron/heliotropes, lasers and other new possibilities.

With improvements in the prospects for attaining critical plasma conditions, extensive, long-term research in advanced reactor engineering technology has also started.

(2) Reactor Core Technology

The plasma confinement systems now proposed for research and development, shown in Table 2-13, the magnetic confinement system and the inertial confinement system, are described below.

Table 2-13. Plasma Confinement Systems Now Under Research

	[System]	[Magnetic field energy]	[Shape]	[Plasma current]	[Type]	[Proposed by]	[Research in Japan]
Nuclear fusion	Magnetic field confinement system	*Low β type	Open		Tandem mirror	US and USSR	Tsukuba University
			Torus	[Uses plasma current for confinement]	Tokamak	USSR	Japan Atomic Energy Research Institute, and Nagoya University
				[Does not use plasma current for confinement]	Stellarator	United States	Nagoya University
					Bumpy Torus	United States	〃
					Heliotron	Japan	Kyoto University
		*High β type	Open		Linear pinch	Unites States and others	Nagoya University
			Torus		Toroidal pinch	Unites States and others	Electrotechnical Laboratory
	Inertial confinement system				Laser system		Osaka University
					Electron beam system		
					Ionic beam system		

Note: β represents the ratio of kinetic energy in plasma to the magnetic energy used for its confinement. When this ratio is less than 0.1, it is referred to as a low β-type, and when more than 0.1 but less than 1, a high β-type. With inertial confinement systems using no magnetic field for confinement of plasma, this ratio, the β value, becomes infinite.

Source: The Japan Atomic Energy Research Institute

(a) Magnetic Field Confinement System

This is a system in which plasma is confined to a limited space by means of an electromagnetic force acting between the plasma and a strong magnetic field outside it. Magnetic field confinement systems are classified into many types according to the shapes of the equipment and the principles of the components, as shown in Table 2-13.

Of these types, the present situation of research and development of the Tokamak method is described below. It is the most advanced method in the world, and has shown considerable progress toward accomplishing practical applications.

The Tokamak type shown in Figure 2-14 is a system first proposed by researchers in the USSR. Compared with other magnetic field confinement systems, it is relatively simple in construction, permitting easy preliminary stage experiments with compact equipment. For this reason, the system has become a major means of magnetic field confinement.

Japan has accomplished outstanding results with Tokamak-type devices through its TNT-A, JFT-2 and JFT-2a equipment, and through U.S.-Japan joint research on Doublet III equipment, based on a May, 1979, agreement.

The major accomplishments are shown in Table 2-15. In the data presented in the table, the β value is an index indicating the efficiency of plasma confinement. The larger the β value, the grater the amount of plasma which can be confined by means of a magnetic field of the same intensity, in turn reducing the size of equipment and construction costs. To raise the value of β, a high level of plasma control is necessary. In this sense, the β value of 4.6 percent achieved by the Japanese research team in April, 1982, is a world record, far surpassing the previous β value of 3.5 percent. This indicates that research and development of the Tokamak type in Japan is about to advance beyond the demonstration stage. As can be seen from the same table, although the plasma core temperature has not reached the world level, rapid progress in reactor core technology for the Tokamak type attained in the past several years has brought Japan to an international level as a whole, considering that the average density of plasma has reached the world's highest level.

Besides the Tokamak type, Japan has achieved valuable results in stable confinement of plasma without electricity, in the stetallator system and heliotron system, formation of electrostatic potential in the tandem mirror system, and realization of high plasma, all contributing to the success of the confinement principle and the development of reactor core technology.

Fig. 2-14. Concept of Tokamak-Type Devices

(1) Low β torus-type magnetic field confinement system

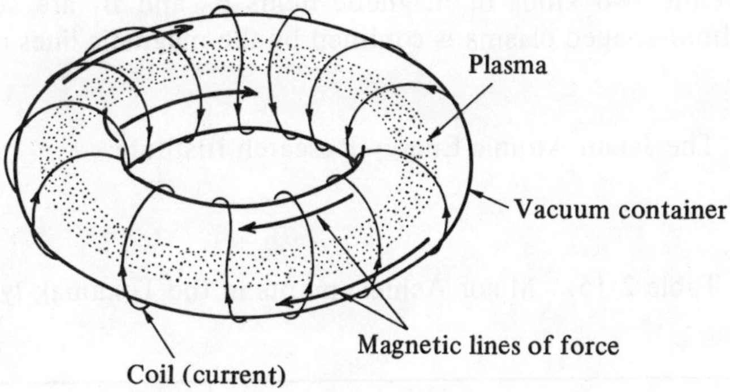

(2) Structure of magnetic field in the Tokamak type confinement system

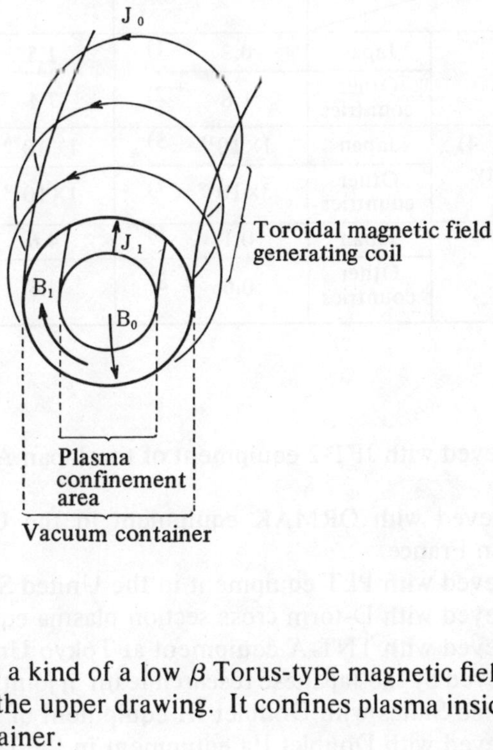

Explanation

1. The Tokamak type is a kind of a low β Torus-type magnetic field confinement system, as shown in the upper drawing. It confines plasma inside a doughnut-shaped vacuum container.
2. The feature that distinguishes the Tokamak type from other low β Torus-types lies in the construction of the magnetic field, as shown in the lower drawing.

3. When a current J_0 is introduced into the toroidal magnetic field generating coil used in the low β Torus-type, a toroidal (circumferential direction) magnetic field is generated. Simultaneously, a plasma current J_1 is introduced in a circumferential direction inside the doughnut-shaped plasma, and a magnetic field B_1 is generated.
4. When the two kinds of magnetic fields B_0 and B_1 are superimposed, the doughnut-shaped plasma is confined by the magnetic lines of force in a spiral form.

Source: The Japan Atomic Energy Research Institute.

Table 2-15. Major Achievements in the Tokamak-type System

Major parameters (Unit) \ Year		Around 1975	By 1982	Target value
Plasma core temperature (KeV)	Japan	0.3 1)	1.5 1)	20
	Other countries	1.8 2)	7.1 3)	
Average 4) plasma density (m^{-3})	Japan	1×10^{19} 5)	1×10^{20} 6)	10^{20}
	Other countries	3×10^{19} 7)	1×10^{20} 6)	
Average value of β (%)	Japan	0.1 1)	4.6 6)	5.0
	Other countries	0.6 8)	4.6 6)	

Notes:
1. Values achieved with JFT-2 equipment of the Japan Atomic Energy Research Institute
2. Values achieved with ORMAK equipment in the United States and TFR equipment in France.
3. Values achieved with PLT equipment in the United States
4. Values achieved with D-form cross section plasma equipment
5. Values achieved with TNT-A equipment at Tokyo University
6. Values achieved by the Japanese research team in joint research between Japan and the United States with Doublet III equipment of the United States
7. Values achieved with Doublet IIa equipment in the United States
8. Values achieved with Pulsator equipment in West Germany

Source: The Japan Atomic Energy Research Institute

(b) Inertial Confinement System

This system requires an extremely high density of plasma compared with other systems. To achieve this, it is necessary to concentrate energy into a small sphere in which fuel materials are enclosed, and compress and heat the energy with laser and other drive units as shown in Figure 2-16. Just before the high pressure and high temperature plasma expands, there is a slight time lag due to inertia, during which the nuclear fusion reaction is attained.

Fig. 2-16. Concept of Inertial Nuclear Fusion Reaction

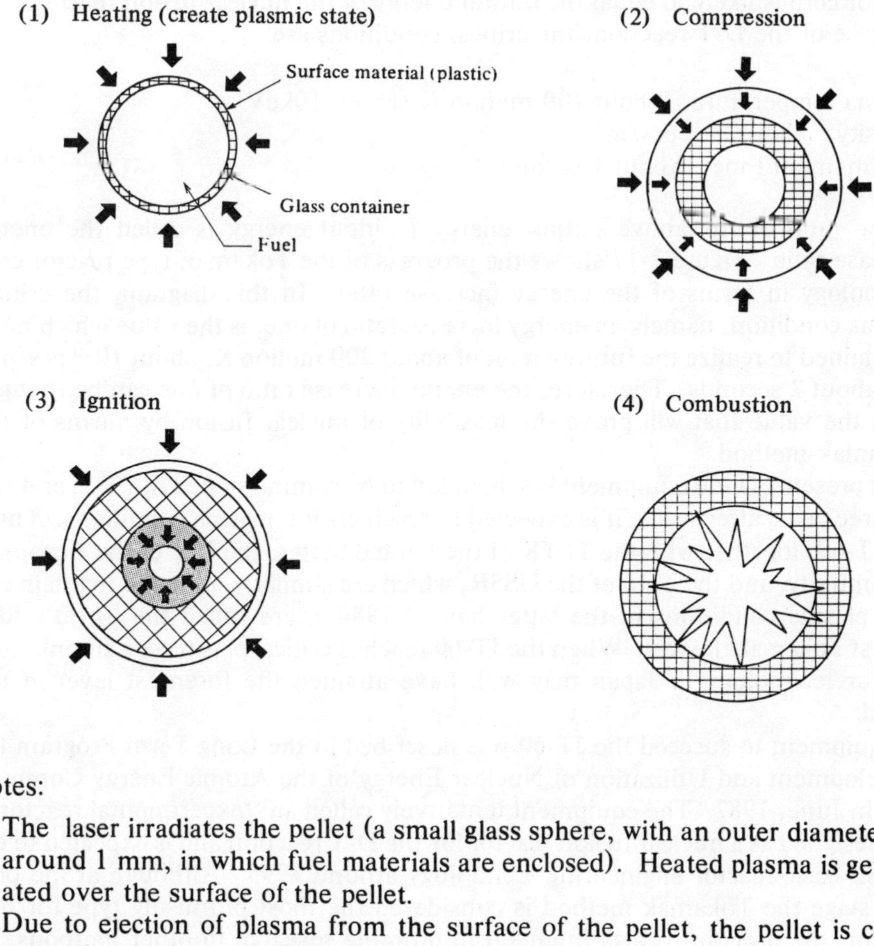

Notes:

1. The laser irradiates the pellet (a small glass sphere, with an outer diameter of around 1 mm, in which fuel materials are enclosed). Heated plasma is generated over the surface of the pellet.
2. Due to ejection of plasma from the surface of the pellet, the pellet is compressed toward the center.

3. The central part of the fuel materials inside the pellet reaches a plasmic state of high temperature and high density, resulting in nuclear fusion.
4. The nuclear fusion reaction consumes all fuel materials, and the energy released is far greater than the energy of the laser need for irradiation.

Source: Osaka University

(c) Future Course of Reactor Core Technology

At present, with construction of the Tokamak experimental equipment JT-60, research on nuclear fusion in Japan has reached the final step toward attaining critical plasma conditions, under which the input energy for heating plasma in the reactor core is likely to equal the output energy of the nuclear fusion reaction. In the case of the D-T reaction, the critical conditions are:

Plasma temperature: About 100 million K (about 10keV)
Density: About 10^{20} pcs/m^3
Confinement time: About 1 second

The ratio of the above output energy to input energy is called the energy increase ratio. Figure 2-17 shows the progress of the Tokamak-type reactor core technology in terms of the energy increase ratio. In this diagram, the critical plasma condition, namely an energy increase ratio of one, is the value which must be attained to realize the future target of about 200 million K, about 10^{20} pcs/m^3, and about 2 seconds. Therefore, the energy increase ratio of one can be thought of as the value that will prove the feasibility of nuclear fusion by means of the Tokamak method.

At present, JT-60 equipment is scheduled to be commissioned in 1985, and two or three years afterwards, it is expected to reach critical plasma conditions. Thus, the JT-60, together with the TFTR of the United States, the JET of the European Community, and the T-15 of the USSR, which are similarly intended to attain critical plasma conditions in the latter half of 1980's, are called the world's four largest Tokamak devices. When the JT-60 reaches critical plasma conditions, core reactor technology in Japan may well have attained the foremost level in the world.

Equipment to succeed the JT-60 was described in the Long Term Program for Development and Utilization of Nuclear Energy of the Atomic Energy Commission in June, 1982. The equipment tentatively called an "experimental reactor," it is designed as a nuclear fusion reactor for the D-T reaction and is expected to establish basic reactor engineering technology around 1995. Although at the present stage the Tokamak method is considered the most promising type for this reactor, the program is also intended to promote research in other methods. If other systems demonstrate the feasibility of attaining critical plasma conditions and can be incorporated into nuclear fusion reactors, then a comparative study with the Tokamak type will be carried out.

Fig. 2-17. Transition of Energy Increase Ratio by the Tokamak Method

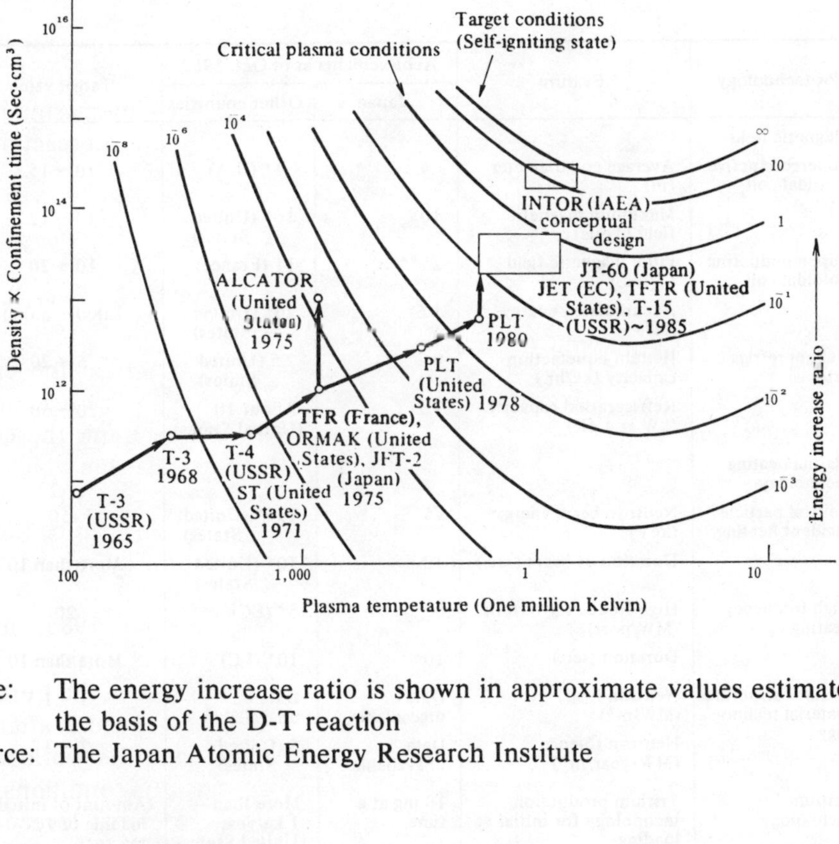

Note: The energy increase ratio is shown in approximate values estimated on the basis of the D-T reaction

Source: The Japan Atomic Energy Research Institute

(3) Reactor Engineering Technology

Full-scale research and development of reactor engineering technology started in the advanced countries several years ago, and has been energetically promoted in line with recent progress in reactor core technology. The major elements of reactor engineering technology are shown in Table 2-18.

First, technology for generation of a magnetic field with a superconductive magnet is indispensable to improving the energy efficiency of the nuclear fusion reactor. In this field, Japan has nearly reached the world level. Moreover, a large scale coil test related to this technology will be carried out in the near future as one of the areas for joint research by the International Energy Agency (IEA), in which one coil will be manufactured in Japan.

Second, plasma heating technology must heat the plasma confined in the magnetic field to the temperature necessary for nuclear fusion reaction. Japanese technology in this field is at the highest level in the world.

Table 2-18. Present Status of Reactor Engineering Technology

Major technology	Feature	Achievements as of Oct. 1982		Target value
		Japan	Other countries	
1. Magnetic field				
Superconductive toroidal coil	Average coil diameter (m)	4	4* (IEA)	10 ~ 15
	Maximum magnetic field (Tesla)	10	10* (United States)	10 ~ 12
Superconducting poloidal coil	Pulse magnetic field (T/sec.)	25**	11 (France)	10 ~ 20
	Accumulated energy (MJ)	1.6	20* (United States)	2.000 ~ 5.000
Helium refrigerator	Helium liquefaction capacity (kℓ/hr.)	0.3	2.5 (United States)	5 ~ 20
	Refrigeration capacity (kW at 4.4K)	1.2	About 10 (United States)	20 ~ 50
2. Plasma heating technology				
Neutral particle incident heating	Neutron beam energy (keV)	75	120* (United States)	150
	Duration of beam (sec.)	10	10* (United States)	More than 10
High frequency heating	High frequency output (MW/port)	5*	5* (EC)	20
	Duration (sec.)	10*	10* (EC)	More than 10
3. Reactor structure material technology	Neutron load (MW/m^2)	Data unavailable	Data unavailable	1 ~ 1.5
	Neutron fluence ($MW \cdot year/m^2$)	Data unavailable	5 (United States)	2 ~ 3
4. Tritium technology	Tritium production technology for initial loading	10 mg at a time	More than 1 kg/year (United States)	(Amount of initial loading of 1 GW reactor) About 10 kg
	Blanket breeding tritium recovery (mg at a time)	About 0.1	10 (United States)	(1 GW reactor) About 100 g/day

Note: * Under manufacture; ** Values achieved using small-scale equipment
Source: The Japan Atomic Energy Research Institute

On the other hand, materials for the reactor itself must be strong enough to withstand exposure to neutrons of high energy generated during the nuclear fusion reaction. Although Japan is at an international level regarding research on materials, it is behind Europe and the United States in research on neutron irradiation, presently being conducted under an agreement between Japan and the United States.

At the same time, tritium technology is indispensable for obtaining fuel when the nuclear fusion reactor operates. Japan, without extensive technology and now about to start systematic research in this field, lags far behind its European and American counterparts.

Also, design is one of the most important aspect of reactor engineering technologies. Although it is difficult to quantitatively compare this technology, Japan is generally considered to be at a high level.

Although reactor engineering technology in Japan has partially reached the international level, on the whole it is a little behind the European and American level, and further efforts are needed.

On the other hand, in reactor core technology, Japan has reached a level comparable to that of Europe and the United States. Considering that the successful accomplishment of the critical plasma conditions with the JT-60 equipment will establish an important milestone toward the realization of the nuclear fusion reactor, further development in reactor engineering technology is expected. The long-term program of the Atomic Energy Commission also stresses that improvements in material technology, technology for production and handling of tritium, remote maintenance technology, and research and development on safety, reliability, durability and stability are all indispensable for the establishment of fundamental reactor engineering technology.

As described above, numerous technological tasks remain to be solved before nuclear fusion can be realized. For Japan to expand research on nuclear fusion as a member of the advanced countries, it will, in principle, have to rely on development of original technology. However, at the same time more international cooperation will also be required in the future in view of the following factors: (1) it will be possible to expand and reinforce the foundations of research by mutually collecting, studying, and utilizing research results obtained in various countries; (2) it will be possible to cut down research and development investment costs and share the risks involved; and (3) it will be possible for Japan to contribute to the progress of the field worldwide.

Space Technology

Ever since the world's first artificial satellite, Sputnik I, was launched 25 years ago, space research has made rapid progress, and to date about 2,800 artificial satellites have been launched, mainly by the United States and the USSR. Today, the utilization of space technology is about to expand into other fields, as illustrated in Figure 2-19. About eighty percent of the artificial satellites launched to date have been for scientific research and have brought about a significant increase in knowledge of space, including the earth itself. In recent years, more practical utilization of satellites, such as for communications and weather observations, has increasingly been carried out.

Space technology requires highly reliable systems that function safely under the harsh conditions in space. Both hardware and software developed for space technology have had significant and far-reaching effects upon all fields of science and technology.

Managed by the Space Activities Commission, space technology in Japan is promoted in compliance with the following fundamental policies: (1) coordination

Fig. 2-19. Utilization of Space Technology

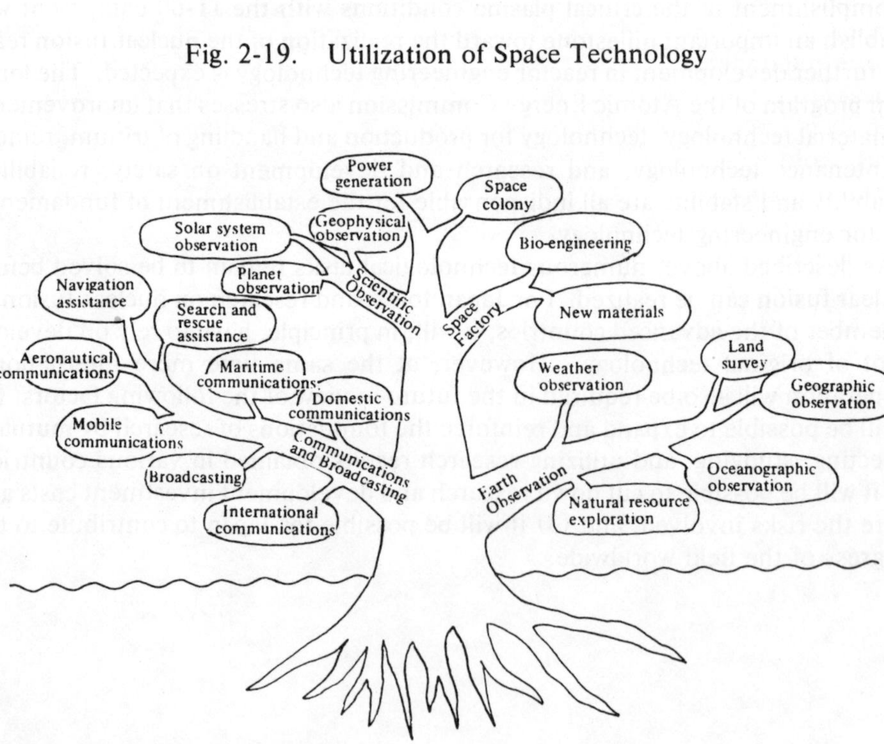

Source: The Science and Technology Agency

with social needs and national power; (2) establishment of independence; and (3) coordination with international activities. The National Space Development Agency of Japan is in charge of development of practical applications, while the Institute of Space and Astronautical Science, set up for joint research by national universities, is in charge of development of scientific research. Moreover, research pertaining to rockets, artificial satellites and their utilization is carried out at national research laboratories.

Space research in Japan began far later than in the United States and the Soviet Union. However, along with the country's increasingly active research and development in artificial satellites in the late 1960's, Japan's first artificial satellite, "OSUMI," was built in 1970. With this, Japan, following the Soviet Union, the United States, and France, became the fourth country in the world to launch an artificial satellite. Having launched twenty-four satellites to date, Japan is fourth in terms of number of launches, but about 96% of the world's total number of launches have been made by the Soviet Union and the United Sates, resulting in a great gap between Japan and these two countries. Even in Japan, however, a number of artificial satellites are scheduled to be launched in the future, as shown in Figure 2-20, and Japan is now about to enter an era of full-scale practical use of space technology.

In order to comprehensively display features of space technology in Japan, exports and imports of space-related equipment are presented in Figure 2-21. As is clear from the diagram, almost all exports are of ground facilities, while the majority of imports are of rockets and aritificial satellites. The fact that Japan produces about one-third of the earth station equipment for INTELSAT (the 106-Nation International Telecommunications Consortium) shows that Japanese technology in this field has reached the world's highest level, but rocket and satellite technology is still lagging.

Fig. 2-20. Artificial Satellites Launched or Scheduled to be Launched in Japan

	Year	1969	1970	1971	1972	1973	1974	1975
Satellites for practical use	N–I Program							ETS-I, ISS
	GMS, CS, BS Programs (Launch commissioned to the United States0							
	N–II Program							
	H–I Program							
Satellites for scientific research		OSUMI	MS-T1	MS-F2	REXS	MS-T2	SRATS	

Already launched   Scheduled

ETS-I : Experimental technology satellite I "KIKU"

ETS-II : Experimental technology satellite II "KIKU 2"

ETS-III : Experimental technology satellite III "KIKU 4"

ETS-IV : Experimental technology satellite IV: "KIKU 3"

ISS : Ionosphere observation satellite "UME"

ISS-b : Ionosphere observation satellite "UME 2"

ECS : Experimental communication satellite "AYAME"

ECS-b : Experimental communication satellite "AYAME 2"

GMS : Geostationary meteorological satellite "HIMAWARI"

GMS-2 : Geostationary meteorological satellite No. 2 "HIMAWARI 2"

GMS-3 : Geostationary meteorological satellite No. 3

CS : Communications satellite No. 2

BS : Experimental medium scale broadcasting satellite "YURI"

Source: The Science and Technology Agecy

1976	1977	1978	1979	1980	1981	1982	1983	1984	1985	1986
ETS-Ⅱ	ISS-b	ECS	ECS-b			ETS-Ⅲ				
	GMS, CS	BS								
				ETS-Ⅳ	GMS-2	CS-2a	CS-2b, BS-2a	GMS-3	BS-2b	MOS-1
									TF	
MS-T3	EXOS-A	EXOS-B, CORSA-b	MS-T4	ASTRO-A		ASTRO-B	EXOS-C		PLANET-A	ASTRO-C

BS-2 : Broadcasting satellite No. 2
MOS-1 : Marine observation satellite No. 1
MS-T1 : Experimental satellite "TANSEI"
MS-F2 : Scientific satellite No. 1 "SHINSEI"
REXS : Scientific satellite No. 2 "DEMPA"
MS-T2 : Experimental satellite "TANSEI 2"
S RATS : Scientific satellite No. 3 "TAIYO"
MS-T3 : Experimental satellite "TANSEI 3"
EXOS-A : Scientific satellite No. 5 "KYOKKO"
EXOS-B : Scientific satellite No. 6 "JIKIKEN"
CORSA-b : Scientific satellite No. 4 "HAKUCHO"
MS-T4 : Experimental satellite "TANSEI 4"
ASTRO-A : Scientific satellite No. 7 "HINOTORI"
ASTRO-B : Scientific satellite No. 8
EXOS-C : Scientific satellite No. 9
PLANT-A : Scientific satellite No. 10
ASTRO-C : Scientific satellite No. 11
TF : Test flight (The second stage payload of TF will be equipped with a land survey apparatus)

Fig. 2-21. Exports and Imports of Equipment Related to Space Development

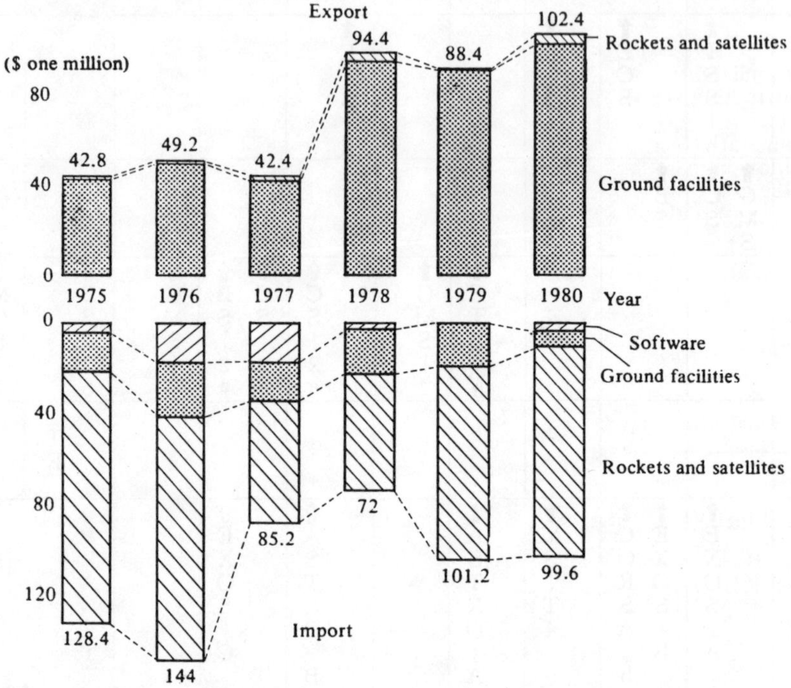

Source: The Society of Japanese Aerospace Companies, Inc.

(1) Rocket Technology

Rockets are classified according to the method by which they obtain thrust. To launch present-day artificial satellites, both liquid fuel and solid fuel rockets are used, with the respective features as shown in Table 2-22. To send satellites into the desired orbit, high precision is required of the rocket. To launch large artificial satellites and geostationary satellites, liquid fuel rockets, with stronger thrust and higher orbiting precision, are mainly used.

In Japan, the Institute of Space and Astronautical Sciences, originally part of Tokyo University, has been carrying out the research and development of solid fuel rockets since its initial experiments with the so-called "Pencil Rockets" in 1955. Since the launching of Japan's first artificial satellite "OSUMI", twelve scientific research satellites have been launched with solid fuel rockets at Kagoshima Space Observation Center. The Science and Technology Agency car-

Table 2-22. Features of Solid and Liquid Fuel Rockets

Solid fuel rockets	Liquid fuel rockets
(Advantages)	(Advantages)
o High reliability because of simple construction and few moving parts	o High performance
o Comparatively easy in development and low in manufacturing cost	o High precision induction control is possible
o In the case of small rockets, the ratio of mass of fuel to rocket is large	o In the case of large rockets, the ratio of mass of fuel to rocket is large
	o Combustion test of an actual rocket engine to be launched is possible
(Disadvantages)	(Disadvantages)
o The specific thrust is comparatively low	o Comparatively difficult in development and manufacture, and expensive
o Difficult to suspend combustion and perform precise induction control	o Liquid fuel is not easy to handle in some case
o The combustion chamber becomes too large and heavy for large-size rockets	

Source: The 1980 Space Development Handbook of the Space Activities Promotion Council of the Federation of Economic Organizations (KEIDANREN)

ried out research and development on both solid and liquid fuel rockets for launching artificial satellites from 1960 until its functions were transferred to the National Space Development Agency of Japan (NASDA) upon its establishment in 1969. While introducing technology from the United Sates, NASDA has also promoted research and development, mainly on liquid fuel rockets. Using the Tanegashima Space Center, NASDA has to date launched nine artificial satellites, including Japan's first geostationary satellite ETS-2 "KIKU-2" using its independently-developed N-I and N-II rockets. In an effort to further enhance rocket performance, NASDA in 1981 commenced development of the H-I rocket, using mostly Japan's own technology for the second and third stage propulsion systems and the induction control system.

Figure 2-23 and Table 2-24 show the major rockets in the world used to launch artificial satellites. The boosting capacity of a rocket is considered an index of its overall performance, and Japanese rockets' boosting capacity are below the international level. However, the H-I rocket, scheduled for test launch in 1985, is capable of launching into geostationary orbit an artificial satellite of about 550kg.

Since ETS-I "KIKU" was launched in 1975, the liquid fuel N-I rocket has been used to launch seven satellites. It has a capacity of launching an artificial satellite of about 130kg into geostationary orbit. The second stage propulsion system for

Fig. 2-23. External View of the World's Major Satellite-Launching Rockets

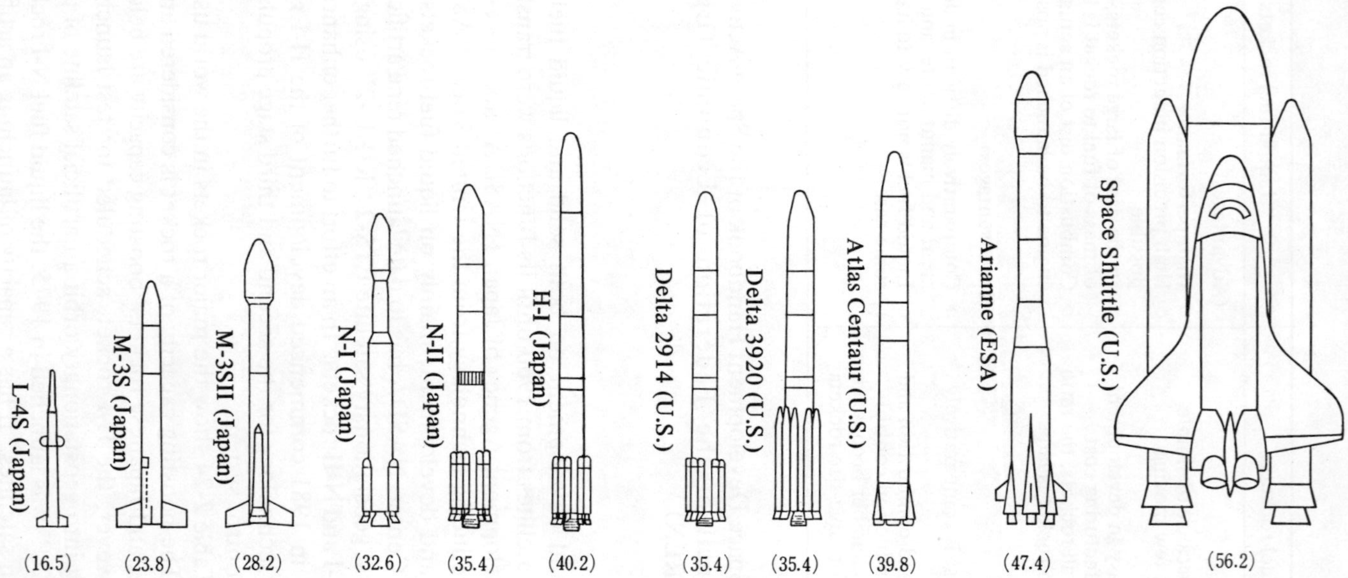

Note: Values in parentheses show overall length in meters
Source: The Science and Technology Agency

Table 2-24. Major Rockets in the World

Rocket	Country	Max. diameter (m)	Overall length (m)	Weight at launching (ton)	Number of stage	1st stage thrust (ton)	Launching capacity	Remarks
Delta 2914	U.S.	2.44	35.36	132	3	234.5	1.769 kg/555 km About 720 kg/geostationary transfer	
Delta 3920/ PAM	U.S.	2.44	35.4	193.5	3	324.0	About 3,400 kg/182 km (3920) 1,270 kg/geostationary transfer	
Atlas Centaur	U.S.	3.05	40.8	148	2	195	970 kg/geostationary transfer	o All stages liquid fuel
Space Shuttle	U.S.	Overall width: 23.8 m Overall height: 23.4 m	56.2	2,041	–		29,500 kg/400 kg	o Repeated use is possible o Manned flight
SSUD-D		1.219	About 1.81	About 2.79	1		635/geostationary transfer	o Upper stage motor for the space shuttle
SSUS-A		About 1.5	About 2.29	About 5.65	1		1,020 kg/geostationary transfer	o Same as above (under development)
IUS		About 2.9	About 5.0	About 16.66	2		2,270 kg/geostationary transfer	o Same as above (under development)
Arianne I	ESA	3.8	47.4	210	3	245	4,500 kg/400 km 970 kg/geostationary transfer	o All stages liquid fuel
Arianne II	ESA	3.8	48.7	214	3	264	1,180 kg/geostationary transfer	o Same as above (under development)
L-4S	Japan	0.735	16.5	9.4	4		About 26 kg/250 km	o All stages solid fuel o Launched Japan's first artificial satellite in 1970
M-3S	Japan	1.41	23.8	49.5	3	223	About 290 kg/250 km	o All stages solid fuel
M-3S II	Japan	1.41	28.2	61	3	197	About 770 kg/250 km	o Same as above (under development)
N-I	Japan	2.44	32.6	90.4	3	149.1	About 1,200 kg/300 km 130 kg/geostationary transfer	o Lanched Japan's first geostationary satellite (ETS-II) in 1977
N-II	Japan	2.44	35.4	134.7	3	291.3	About 2,000 kg/300 km 350 kg/geostationary transfer	
H-I	Japan	2.44	40.3	139.2	3	291.3	About 3,000 kg/300 km 550 kg/geostationary transfer	o (Under development)

Note: Launching capacity shown in weight of satellite/orbit. For example, 1,769kg/555km means capability to launch a 1,769kg satellite to an orbit 555km from earth.

Source: The Science and Technology Agency

the N-I rocket has been developed by the Science and Technology Agency and NASDA. Its engine, the LE-3, used for the second stage propulsion system, has a thrust of 5.44 metric tons and a specific thrust of about 290 seconds in vacuum.

The specific thrust is the most important index of rocket engine performance, and the larger the value, the better the performance. Figure 2-25 shows the specific thrust of various types of liquid fuel engines.

As the figure shows, the advantage of an engine using liquid oxygen as an oxidizing agent and liquid hydrogen as a fuel (the liquid oxygen — liquid hydrogen engine) is that it has a high specific thrust. This type is considered to be promising as a liquid fuel engine for large-scale rockets, and has been used for the main engine of the U.S. space shuttle. Therefore, joint research on this engine was started between the National Aerospace Laboratory of the Science and Technology Agency and the National Space Development Agency of Japan in the early 1970's, as well as at the Institute of Space and Astronautical Sciences. The results of this research have been incorporated into the LE-5 engine, scheduled to be used for the second propulsion system of the H-I rocket under development by

Fig. 2-25. Specific Thrust of Various Liquid Fuel Rocket Engines

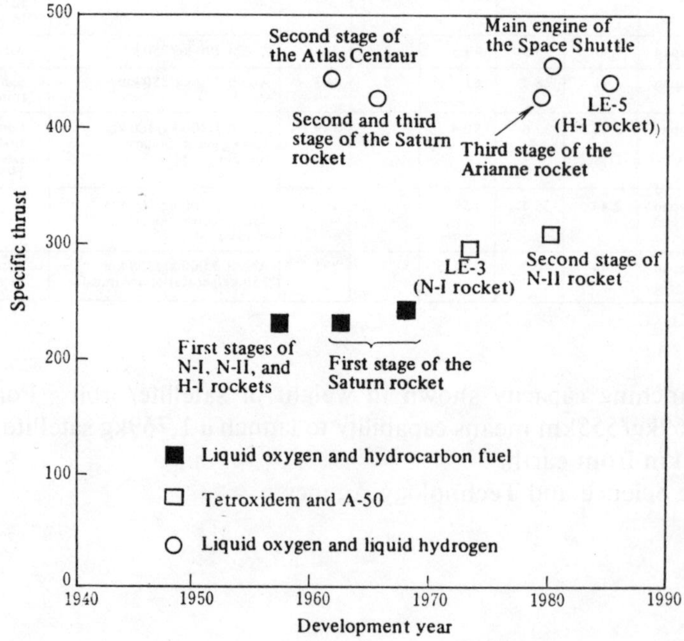

Source: The National Space Development Agency of Japan (NASDA)

the National Space Development Agency of Japan in preparation for the launching of large-scale satellites around 1985. In combustion tests in 1980 and 1981 on the prototype engine of an experimental-model of the LE-5, the required engine performance was achieved. In 1982, the combustion tests were started on a production engine equivalent to that to be used in the actual rocket. Research on the technology of elements required to improve combustion pressure and performance of this engine has been conducted by the National Aerospace Laboratory of the Science and Technology Agency, and results of this research are expected to contribute to the improvement of rocket performance in the future.

The induction control system of a rocket maintains a rocket stablity and controls the direction and speed needed to place a satellite into a specified orbit. For this purpose, the induction control system has the following functions:

(1) Monitoring the position and speed of a rocket
(2) Computing the flight direction and speed required to reach the targeted position and speed on the basis of the above information
(3) Changing the direction of injection and the combustion time of the rocket engine in accordance with the above calculations

The first two functions are called induction, and the third, control. Induction systems can be roughly classified into radio induction and inertial induction systems.

In the N-I rocket, induction is carried out from the ground using a radio induction system, as shown in Figure 2-26. The equipment mounted on the rocket is comparatively simple, but it is possible to control the rocket only when it is within the reach of radio waves from the earth station (visible range). Since the rocket is controlled according to the programmed induction system built in the rocket, its orbital precision is not so high. Therefore, the inertial induction system has been adopted for the N-II and H-I rockets.

The inertial induction system mounted on the rocket detects changes in acceleration and position using an accelerometer and gyrocompass, and computes corrections in the direction and speed of the rocket with a computer, making it unnecessary to control the rocket from the ground. However, the equipment on the rocket must have far higher performance than radio induction system equipment, as well as ability to withstand severe flight conditions. Although inertial induction equipment developed for the U.S. Delta rocket has been used in the N-II rocket, comparable equipment for the H-I rocket is now being developed independently at the National Space Development Agency.

Fig. 2-26. Concept of Radio Induction System

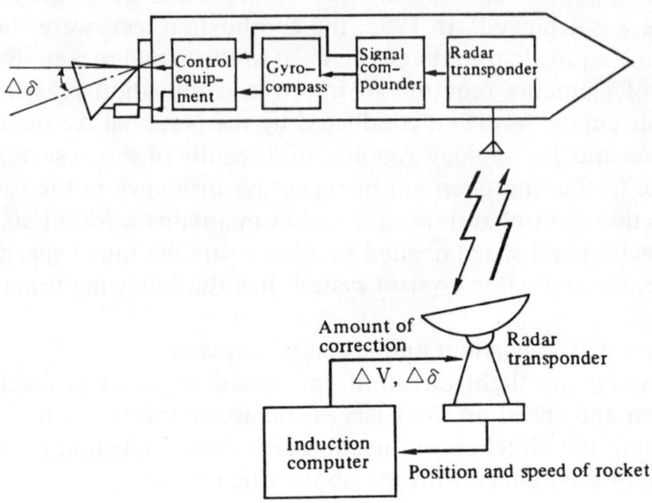

Note: ΔV: Amount of speed correction

$\Delta\delta$: Amount of directional correction

Source: Same as Table 2-22.

(2) Artificial Satellite Technology

Although an artificial satellite is composed of several sub-systems, as shown in Table 2-27, they can be roughly classified as the payload for attaining the respective objectives and the bus system for obtaining stable operation of the payload as well as for protecting it from the external environment. The artificial satellite must perform certain missions such as communications, observations, or scientific research, and operate stably for a predetermined period in a severe environment. Consequently, for development of an artificial satellite, overall technological capability, including design, assembly, and tests, is required in addition to high-level pioneering technology.

Because Japan started development of artificial satellites for practical applications later than the United Sates, the Soviet Union and Western Europe, the country has been absorbing technology from the United States in an effort to promote growth of its own technology, and year by year an increasingly larger portion of satellite components have been domestically manufactured. For example, the CS (the experimental, medium-size fixed communication satellite "SAKURA" launched in 1977) was developed substantially with technology from the United States. In contrast, the CS-2 (the communication satellite No.2), now under

Table 2-27. Sub-System of an Aritificial Satellite

	Equipment/System	Outline
Payload	Sub-system for attaining the objective	Equipment needed to attain the objectives of satellites, such as relays and antennas in communication satellites, or visual and infrared scanning radio-meters in meterorological satellites
Bass System	Sub-system for position control	Detects and controls position of the satellite to direct antennas and observation systems toward an intended direction
	Sub-system for power supply	Supplys electric power to respective parts of satellites
	Sub-system for telemetry and command	Transmits data to earth and receives commands from ground stations
	Sub-system for propulsion control	Changes or corrects the orbit and position of satellites
	Sub-system for heat control	Controls temperature inside satellites
	Outer shield	Protects satellites from heat and shock

Source: The Science and Technology Agency

development and scheduled for launching in 1983, has a capacity and performance roughly equivalent to the CS, but with more than sixty percent of the components manufactured, assembled and tested in Japan. In this way, Japan has steadily built on its achievements and developed its own technology to the point where it can contribute to INTELSAT satellites.

The attitude of an artificial satellite, or the inclination of its axes relative to the earth, always fluctuates in orbit due to the effect of heliotropic winds, the specific gravity of the earth, terrestrial magnetism, etc. The fluctuation of the attitude causes the following problems:

(1) Observation equipment and antennas on the satellite get out of line from those programmed.
(2) The surface of the solar-cell supplying electric power to the satellite does not efficiently receive the solar rays.
(3) Corrections in the orbit of satellite become increasingly difficult.
(4) Temperatures inside the satellite fluctuate widely.

For these reasons, it is essential to control the attitude of the artificial satellite, which, as shown in Table 2-28, is done by two systems, a spin stabilizing system and a three-axial system. Figure 2-29 shows external views of representative satellites using each of these systems. The spin stabilizing system has been adopted for a number of artificial satellites because of its simple principle.

The three-axial attitude control system is a comparatively new system applied for the first time in the geostationary satellite ATS-6 (Applications Technology Satellite No.6) of the United States, launched in 1974. Although this system has not been adopted widely in Japan, internationally it has in recent years increasingly been adopted because it permits mounting of large solar-cell panels and complex antenna systems. Especially in Europe, this system has in recent years been adopted in the majority of satellites.

In Japan, the three-axial attitude control system is used in the BS "YURI," launched in 1978; the ETS-III "KIKU," launched in September, 1982; the BS-2 (Broadcast Satellite No.2); and the MOS-1 (Marine Observation Satellite No.1), now under development.

Since the three-axial attitude control system requires an extremely high degree of technology, further research is needed for large-scale satellites of the future.

Table 2-28. Comparison of Attitude Control Systems

System		Principle	Characteristics
Spin stabilizing system		Stabilized by rotating a satellite like a top	o The mechanism is comparatively simple o The stability of the axis of rotation is high o It is difficult to obtain large electric power due to limited size of solar cell panels that can be mounted o The design of satellites is limited
Three-axial system	Zero momentum system	Stabilized by using three fly wheels that absorb the disturbing force	o The mechanism is comparatively complex o It is possible to obtain large electric power using large solar cell panels o The design of satellites is unlimited o Generally, the consumption of energy is larger than spin stabilizing satellites
	Bias momentum	Stabilized by rotating a large wheel inside a satellite	

Source: The Science and Technology Agency

Fig. 2-29. External View of Satellites Using Either Spin Stabilizing or Three-Axial Attitude Control System

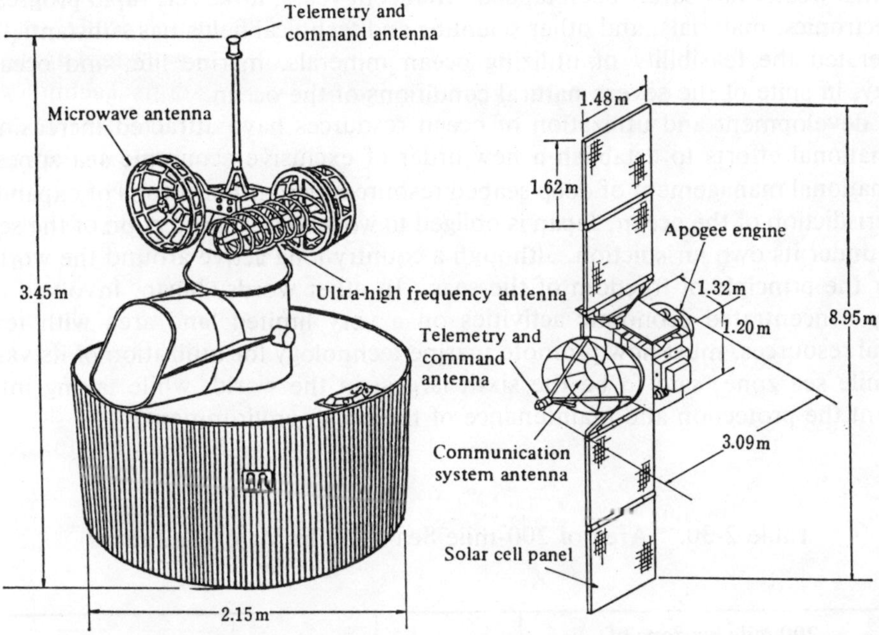

(a) Spin stabilizing system (GMS-2 "HIMAWARI 2")

(b) Three-axial system (BS "YURI")

Source: The National Space Dvelopment Agency of Japan

Direct research as well as research on related technology (such as the gyrocompass, wheel, large-scale solar panels) is being promoted by both the National Aerospace Laboratory of the Science and Technology Agency and the National Space Development Agency of Japan.

As described above, Japan, which started space development later than Europe and the United States, is achieving results in the fields of both rocket and artificial satellite technology, but is still largely dependent on technology introduced from other countries. Because space technology not only exerts a significant effect upon the development of natural science in general but also contributes to improvements in the living environment in such fields as communications, broadcasting, meteorological and other observations, all of which bring economic benefits, Japan needs to actively promote research and development with a view to developing original technology, and take an even more positive part in international cooperation in space development in the future.

Marine Technology

Although the ocean has been used for centuries for fishing and shipping, its potential wealth has barely been tapped. In recent years, however, rapid progress in electronics, materials, and other scientific and technical fields has substantially accelerated the feasibility of utilizing ocean minerals, marine life, and ocean energy, in spite of the severe natural conditions of the ocean.

As development and utilization of ocean resources have attracted increasing international efforts to establish a new order of exclusive economic sea zones, international management of deep seabed resources, and other means of expanding jurisdiction of the ocean, Japan is obliged to work toward protection of the sea zone under its own jurisdiction, although a country long active around the world under the principle of freedom of the seas. In other words, Japan, involved in highly concentrated economic activities on a very limited land area with few natural resources, must now promote marine technology for utilization of its vast 200-mile sea zone, said to be the sixth largest in the world, while taking into account the protection and maintenance of the ocean environment.

Table 2-30. Area of 200-mile Sea Zone in Major Countries

200-mile sea zone of major countries			Land area	Total land and sea area	Ratio of sea area to land area
Order	Country	Area			
1	United States	762	936	1,698	0.8
2	Australia	701	769	1,470	0.9
3	Indonesia	541	190	731	2.9
4	New Zealand	483	27	510	17.9
5	Canada	470	998	1,468	0.5
6	Japan	451	38	489	11.9
7	Soviet Union	449	2,240	2,689	0.2
8	Brazil	317	851	1,168	0.4
9	Mexico	285	197	482	1.5
10	Chile	229	76	305	3.0
:					
24	England	64	24	118	3.9
45	France	34	55	89	0.6
96	West Germany	4	25	29	0.2

Source: The U.S. Department of State and others

Unlike land, the ocean is an environment of large fluctuations due to tidal and ocean currents, subject to corrosion, high pressure and other restrictions caused by sea water, with the added difficulties of transmission of light and electromagnetic waves. Marine science and technology must develop comprehensive technology that systematically integrates such elements as special materials and structures meeting the severe requirements of the various natural conditions of the ocean.

(1) Diving Technology

In the past, diving was originally intended for fishing or as a kind of sport, and the maximum depth was around 20 to 30m. However, as man's knowledge expands and the possibilities of utilization of the sea are increasingly recognized, man has become interested in scientific investigation of the sea for such activities as underwater petroleum resources, mainly on the continental shelf, and development of fishery resources by establishing artificial fishing grounds. To construct and maintain underwater structures required for these facilities, men must dive into the sea.

However, since the pressure of the sea is raised by as much as 1 atm for every 10 meters, divers are subjected to extensive physical stress, and even for a person accustomed to diving without any diving apparatus, at 30 to 40 meters in depth activity is substantially limited to two or three minutes. Diving was popularized after 1943, when the Frenchman Cousteau succeeded in developing an underwater breathing apparatus, called the "aqualung". Because of his success, France has become a leading country in ocean development. However, since compressed air was initially used in the aqualung, at or below 60m nitrogen in the compressed air dissoloved in the diver's body and caused central paralysis, or what is called critical nitrogen sickness. In the latter half of the 1940's, nitrogen was replaced by helium in order to prevent nitrogen sickness and shorten the pressure reducing time, bring into existence a helium-oxygen breathing apparatus which made it possible to dive deeper than when using compressed air. Then, in 1960, saturation diving technology was developed, making it possible to dive safely for long hours without any physiological trouble even after the mixture of helium and oxygen dissolved in the body has reached the saturation point.

For development of saturation diving technology, a high degree of science and technology in a number of fields such as medicine and physiology is necessary. Progress in diving technology drastically increased record diving depths for human beings, as shown in Figure 2-31, to about 500m in actual sea diving and about 700m in simulation tests in high pressure tanks.

Although France pioneered research and development of diving technology and has been one step ahead of other countries in this field, the United States and England have also been conducting full-scale research and development for petroleum exploration and military purposes. In Japan, a country poor in natural resources and energy, the importance of the development of natural resources on the continental shelf has been apparent since the beginning of ocean exploration, and technology related to diving work has been needed. However, partly because research and development of such technology commenced comparatively later than in other countries, the Japanese technological level in this field is significantly behind the international level. Technology for diving to a depth of 100m in actual sea conditions was developed under the Seatopia project from 1968 to 1975, and in 1979, based on the results of this and other research, an experimental simulation diving equivalent to 300m was achieved. This experiment, using the

Fig. 2-31. Major Saturation Diving Experiments

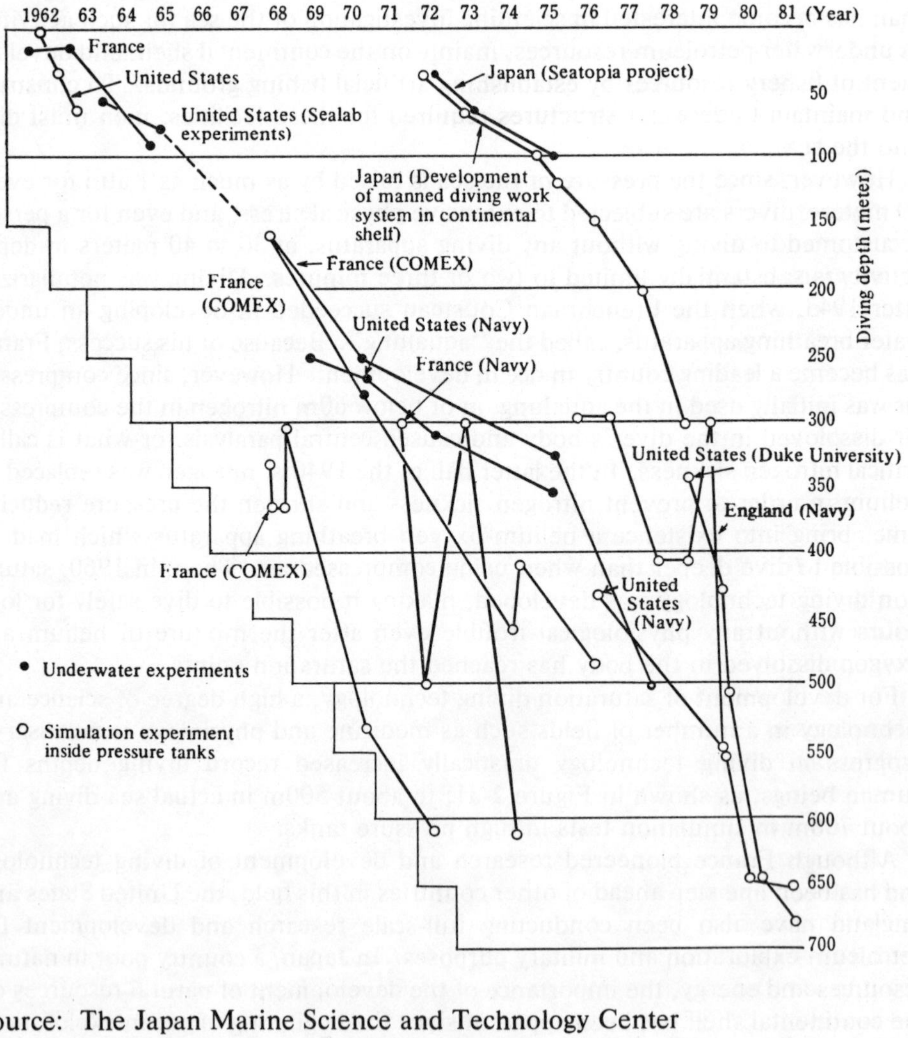

Source: The Japan Marine Science and Technology Center

principle of saturation diving technology, was carried out at the Japan Marine Science and Technology Center for the purpose of developing a diving system which would enable work underwater to be safely and efficiently performed for long periods of time. At the same time, research on construction of a deck decompression chamber, underwater elevator, submarine work base, supporting buoy, and related facilities was carried out as part of research on a submarine working system.

In the future, Japan plans to improve diving technology by conducting diving experiments at depths of 300m and to develop practical diving technology workable at depths of up to 100m. At the same time, the country needs to develop various kinds of diving apparatuses and to improve the safety and efficiency of the system.

(2) Deep Sea Submarine Technology

To investigate conditions of the deep sea directly with the human eye, manned oceanographic submarines are needed. This will make it possible to carry out research on manganese nodules and other abyssal floor mineral resources, abyssal life, and underwater geological structures.

However, the abyssal environment contains high pressure, darkness, low temperature and other severe conditions. Therefore, building of a deep sea oceanographic submarine requires development of methods of manufacturing a pressure resistant hull, including materials which can withstand extremely high pressures; buoyant materials with a low specific gravity to enhance mobility by making the hull compact in size and light in weight; a high pressure pump to take on and discharge sea water when diving and surfacing; and ultrasonic wave technology for communication with the mother vessel on the surface, as well as other high level technology used in the abyssal environment. Figure 2-32 shows the submersion depths of major oceanographic submarines which have so far been developed or are under development. The United States and France are world leaders in this field because of their active promotion of research and development.

Recently, to replace the bathysphere-type submarine, which has poor mobility, development of a compact and highly mobile oceanographic submarine is being promoted in order to permit a wider range of investigations. Examples of such submarines are the ALBIN of the United States, submersible up to a depth of 4,000m, and the (CYANA) of France, submersible up to a depth of 3,000m. These two countries have the world's foremost technology in building oceanographic submarines, and in 1984, are scheduled to complete construction of submarines submersible up to a depth of 6,000m. In Japan, abyssal exploration was conducted using the KUROSHIO, built in 1960, the SHINKAI, built in 1968, and other oceanographic submarines. With the SHINKAI 2000, completed in 1981, full-scale development of oceanographic submarines began in Japan, enabling it to join the pioneering group of countries which includes the United States, the Soviet Union, France, Canada and others.

Fig. 2-32. Development of Oceanographic Submarines in the World

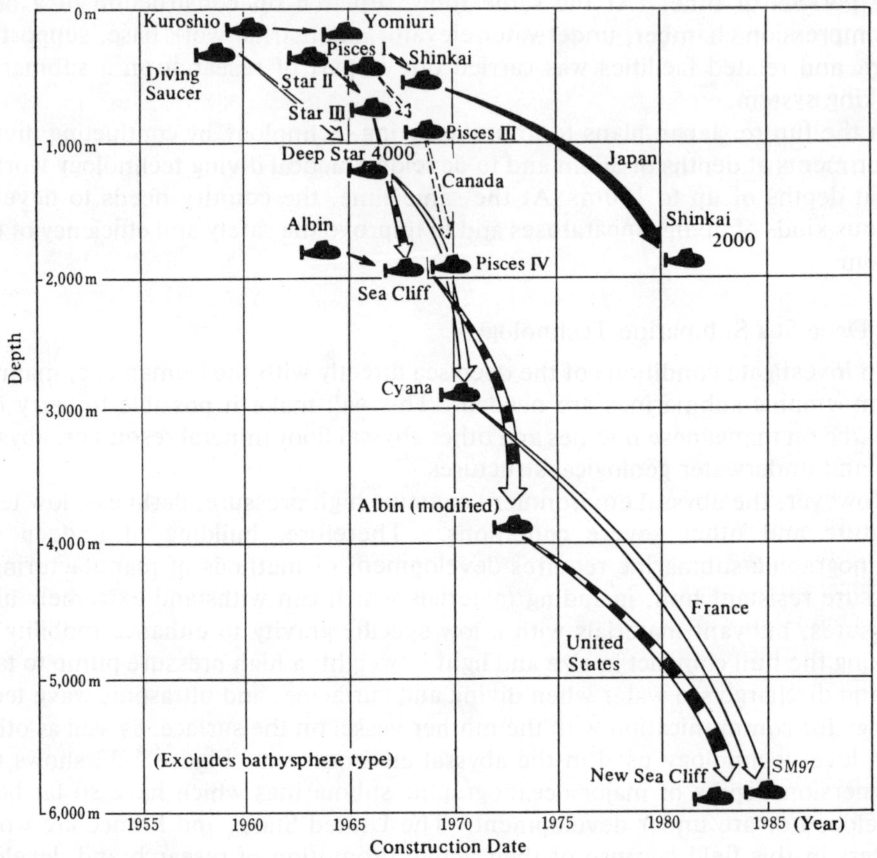

Source: The Japan Marine and Science Technology Center

The SHINKAI 2000 was built almost entirely with Japan's own technology. The inside diameter of its pressure resistant hull, 2.2m, makes it larger than other 2,000m class oceanographic submarines, affording a larger working area. Its advanced systems include a silver-zinc battery high in weight and volume efficiency as a power source. However, with a 2,000m class oceanographic submarine, it is possible to investigate only 16 percent of the seabed in the world, or only around 30 percent of the 200-mile sea zone around Japan. For this reason, development of oceanographic submarines or other systems that permit investigations in far deeper abyssal areas is expected in the future.

In the future, Japan needs to explore manganese nodules, which exist in large quantities on the 4,000 to 6,000m deep abyssal floor and which have been attracting the attention of various countries as a sea resource; metallic sulfide resources recently found in some parts of the world, mainly the Pacific, and said to be worth as much or more than manganese nodules; and abyssal geography earthquake prediction.

For these purposes, a 6,000m class oceanographic submarine, capable of investigating any abyssal area in the world, is needed and must be developed by extending the achievements of the SHINKAI 2000.

Life Science

Life science seeks to clarify biotic phenomena by making use of a wide range of knowledge in biology, medical science, agriculture, chemistry, physics, engineering and other fields. Research in life science, from clarification of biotic phenomena to applications at the level of molecules, cells, tissues, organs, individuals, and populations, is expected to contribute greatly to human welfare through expansion of knowledge of life processes, progress in medical treatment, and solutions to such problems as food and energy supplies.

From among current areas of research in life science, this section will focus on technology for DNA recombination, now considered very significant for its clarification of genetic phenomena and for the promising new technology expected from it in the future.

As illustrated in Figure 2-33, recombinant DNA technology uses such things as enzymes to recombine molecules from multipliable and other kinds of DNA in certain germ cells, and then transfers the molecules to other germ cells. DNA (deoxyribonucleic acid), an organic compound of extremely high molecular weight found in cells, is made up of carbon, oxygen, hydrogen, nitrogen and phosphorus.

Fig. 2-33. Concept of Recombinant DNA Technology

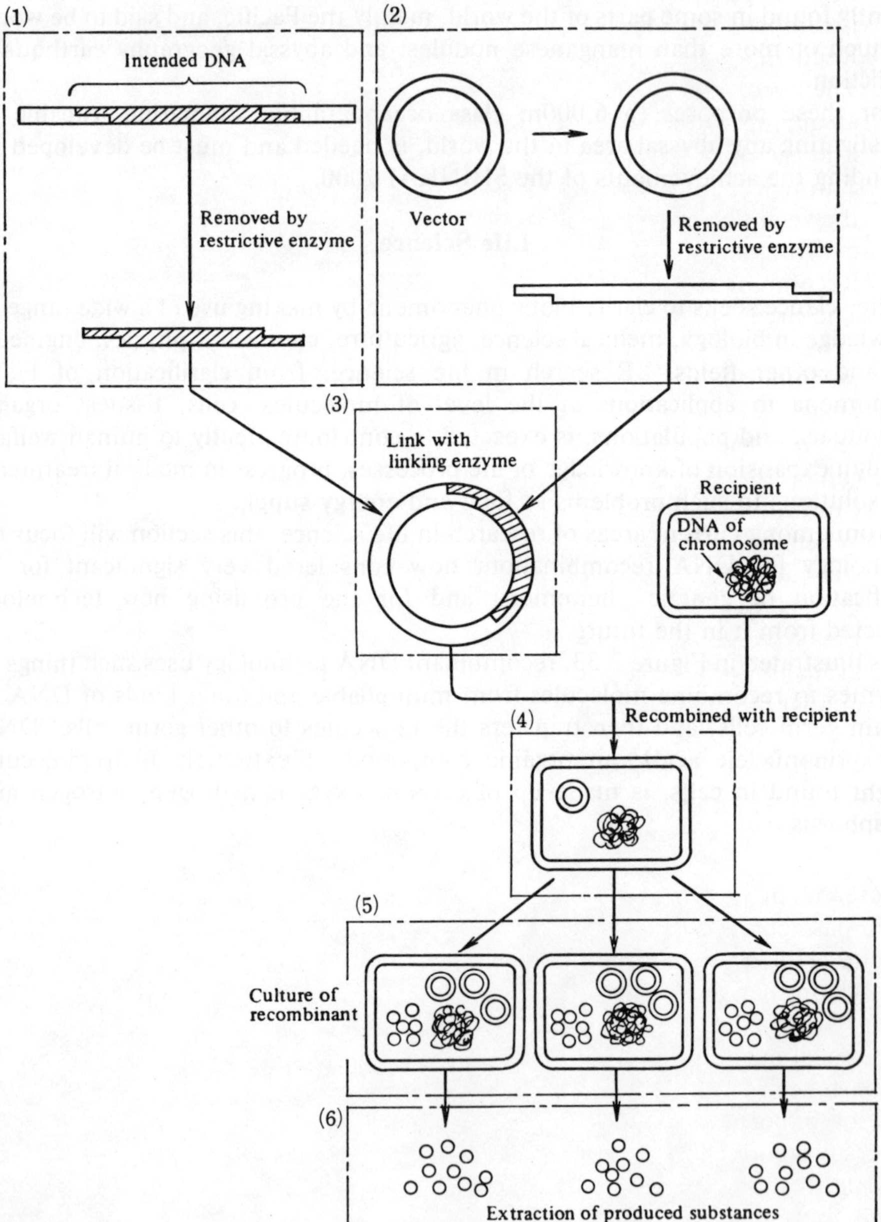

Explanation

(1) Preparation of DNA plug:
A DNA plug, containing information for synthesizing a particular protein, is removed from the DNA in a germ cell by use of a restrictive enzyme.

(2) Regulation of vector:
The circular DNA outside the nucleus, called the vector, is removed from an appropriate micro-organism using the same restrictive enzyme as in (1).

(3) Linkage of DNA plug and vector:
Through the previous two operations, the ends of both the DNA plug and the vector have properties allowing them to be linked together with a linking enzyme, and a new circular DNA (recombinant molecule) is created.

(4) Recombination of circular DNA into recipient:
The circular DNA prepared in (3) is recombined into an appropriate micro-organism recipient, called the "recombinant". Through this operation, the recombinant synthesizes protein according to the information stored in the DNA. Depending on the combination of the recipient and vector, the recombined circular DNA sometimes self-multiplies independently from cell division in its recombinant.

(5) Culture of recombinant:
When the recombinant is cultured under appropriate conditions, it synthesizes protein in compliance with information in the DNA.

(6) Extraction of produced substances:
Finally, desired substances are extracted from the recombinant, and are available for use in a wide range of fields such as medicine, agriculture, animal husbandry, and environmental protection.

Source: The Institute of Physical and Chemical Research

Research on recombinant DNA has been initiated and led by the United States, which has achieved the majority of breakthroughs in producing various kinds of protein. However, in recent years other countries have also made progress in research on such technology, including some important accomplishments made in Japan. Table 2-34 summarizes recent international developments in the field.

The most outstanding features of research on recombinant DNA in the U.S. is the existence of venture businesses organized mainly for research and development of biotechnology with industrial applications, in cooperation with both university laboratories and chemical companies. Table 2-35 shows the projects for mass cultivation of recombinats approved by the U.S. National Institute of Health, indicating that recombinant DNA technology in the United States has already left the laboratory stage and entered the industrialization stage.

Table 2-34. Recent Developments in Recombinant DNA Technology

Year	Countries	Outline
1968	United States	Isolated a restrictive enzyme
1972	United States	Discovered a linking enzyme
1973	United States	Succeeded in the experiment of recombinant DNA for the first time
1977	United States	Succeeded in recombining DNA of rat's insulin with colon bacillus
	United States	Succeeded in making somatostatin by recombining the DNA of artificial synthesized somatostatin with colon bacillus
1978	United States	Succeeded in making insulin by recombining DNA of artificially synthezized insulin with colon bacillus
1979	United States	Succeeded in making growth hormone by recombining DNA of artificially synthesized growth hormone with colon bacillus
1980	United States and Switzerland	Succeeded in making interferon by recombining DNA of interferon with colon bacillus
	Japan	Succeeded in making soybean protein using colon bacillus
	Japan and United States	Succeeded in high efficiency culture of interferon
1981	United States	Succeeded in making interferon using yeast
	Japan	Succeeded in recombining DNA of thermoduric enzyme with colon bacillus
	United States	Succeeded in making interferon using depoliant
	Japan	Succeeded in making secretin using colon bacillus
	United States	Succeeded in making protein of hepatitis B virus using yeast
	Japan	Succeeded in making glutathione using colon bacillus
1982	Japan	Succeeded in making rennin of calfs using colon bacillus
	Japan	Succeeded in efficient culture of protein of hepatitis B virus

Source: The Institute of Physical and Chemical Research

Research on recombinant DNA in various European countries, often promoted as a government project, is said to be one step behind the level of the United States. In England and France, venture businesses in this field have been established through cooperation between government and private enterprise. In West Germany, the European Molecular Biology Organization was established through joint investment of various European countries. Thirteen European scientists

Table 2-35. Private Enterprise Projects for Mass Cultivation of Recombinates Approved by the U.S. National Institute of Health (as of Oct., 1981)

Produced substances \ Enterprises	Company A	Company B	Company C	Company D	Company E	Company F	Company G
Human Insuline	○	○					
Human Somatostatin		○					
Human Growth hormone		○					
Thymosin Alpha		○					
Human Leucocyte Interferon		○	○	○	○	○	
Human Fibroblast Interferon		○		○		○	
Bovine Growth Hormone		○					○
Human Calcitonin		○					
Porcine Growth Hormone		○					
VP_3 Protein of Foot-and-Mouth Disease Virus		○					
Reverse Transcriptase of Avian Sarcoma Virus							○

Note: Mass cultivation refers to production capacity of 10 litres or more.
Source: Prepared by the Science and Technology Agency based on data of the NIH of the United States.

were awarded the Nobel Prize for studies related to life sciences from 1970 through 1982, compared to twenty Americans. Therefore, the European countries are said to have the potential to catch up with the United States in the future.

In Japan, the Council for Science and Technology first called attention to the importance of the research on recombinant DNA in a report issued in May, 1971. In November, 1978, the Science Council under the Ministry of Education proposed procedures to promote experiments on recombinant DNA at universities and other research laboratories, in response to which the Ministry issued guidelines for such experiments in March, 1979. In the same year, the Council proposed safety guidelines for experiments, consisting of two elements: (1) using an apparatus to physically prevent leakage of recombinants from the laboratory; and (2) using recipients that cannot survive outside the laboratory. In August, 1980, the Council proposed research targets and promotion policies stressing the importance of research in bio-technology.

Currently, a variety of research and development is being conducted at universities and national laboratories, as well as at a number of leading enterprises working on industrialization of recombinant DNA technology, at present mainly to increase the quality and value of pharmaceutical products, as shown in Table 2-36.

Table 2-36. Plans for Industrialization of Recombinant DNA Technology by Private Enterprises in Japan

Industrial classification (Number of enterprises answered) \ Application / Main products	Pharmaceutical	Chemical and food	Cultivation of plants	Total
	Antibiotics, interferon, enzyme for medical use, medical reagents, etc.	Amino acid, agricultural chemicals, yeast, etc.	Barley, etc.	
Pharmaceuticals (12 companies)	6	–	–	6
Chemicals (27 companies)	10	8	1	14
Textile, paper and pulp (12 companies)	4	–	1	4
Food (24 companies)	6	9	1	11
Others (10 companies)	–	1	–	1
Total (85 companies)	26	18	3	36

Notes: 1. Number of enterprises planning to industrialize the recombinant DNA technology.
2. Since some enterprises are planning applications in two or more fields, sub-totals and grand totals do not match in some cases.
3. Figures in parentheses refer to the number of enterprises planning to industrialize bio-technology, including recombinant DNA technology.

Source: A survey conducted in June, 1981, by the Nippon Keizai Shinbun, Inc.

As shown in Table 2-37, of 55 patent applications related to recombinant DNA technology made in Japan as of May, 1981, American applications amounted to 27, or 49 percent, while Japanese applications amounted to 12, or 22 percent. Considering that the applications from France, England, West Germany and other foreign countries amounted to 16, or 29 percent, it can be seen that Japan is below the international level.

Although a late-starter, Japan has a tradition of micro-organism utilization in soy paste, soy sauce and other fermented products, and utilization of technology for the manufacture of antibiotics, amino acids, etc. has in recent years reached the highest level in the world.

For this and other reasons, Japan is expected to play a significant role in international competition in recombinant DNA technology. For example, production in Japan of the hepatitis B virus protein has reached about 500,000 molecules per spore of yeast, the highest value in the world, far surpassing previous results in the United States, paving the way for application of recombinant DNA technology to the manufacture of vaccines. This shows that research on recombinant DNA in Japan is gradually reaching the international level.

In November, 1981, the government submitted to the Council for Science and Technology an important proposal concerning a basic program for development of pioneering technology in life science, indicating the future direction of research in bio-technology in Japan. The proposal, now under deliberation, is scheduled for evaluation in the first half of 1983.

Recipients used in Japan are generally limited to colon bacillus and four other strains. Although guidelines for experiments have been partially revised, restrictions are still numerous when compared with other countries, particularly the United States. The Council of Science and Technology indicated that the range of these recipients must be expanded for research to progress, but it is first necessary to know what changes will occur in various recipients when different kinds of DNA are recombined. For this reason, the council also pointed out the need to establish laboratories where research on safety aspects can be conducted.

Fig. 2-37. Number of Recombinant DNA Technology Patent Applications Applied for in Japan (Cumulative)

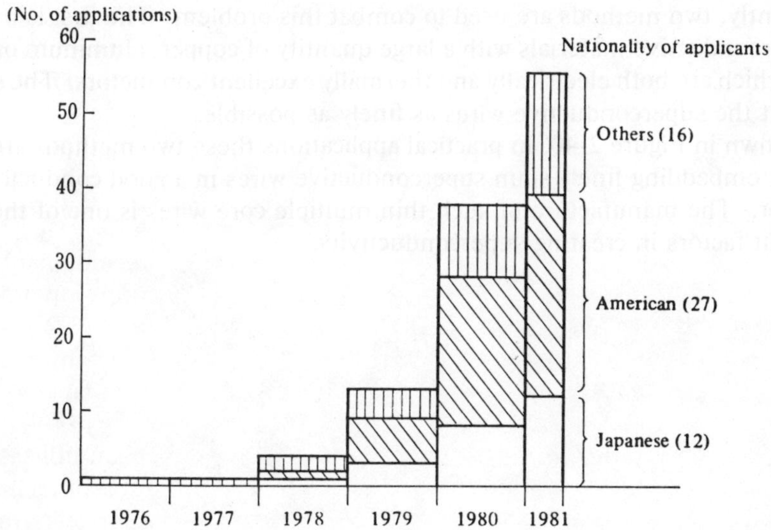

Source: The Japan Institute of Invention and Innovation

Processes Under Extreme Conditions

Along with the remarkable progress of technology as well as social and economic changes, the need to develop new materials and high level systems has become increasingly apparent. Accordingly, technology under extreme conditions such as very low temperature, super-high pressure, extremely high temperature and ultra-high vacuum, opens up new possibilities in super conductivity, synthesis of new materials, and so on. Of these, this section will discuss progress in technology using extremely low temperatures and super-high pressures.

(1) Technology Under Extremely Low Temperatures

An extremely low temperature is usually defined as one lower than that of liquified natural gas (-160°C). This is usually used for creating superconductivity, which is expected to be utilized in a variety of fields as shown in Figure 2-38, and to have far-reaching effects, several of which will be discussed below.

(a) Superconductive Materials

Superconductivity occurrs under conditions determined by temperature (T), electric current density (J), and magnetic field (H), which differ according to the material, as shown in Figure 2-39. The higher these values are, the greater the superconductivity of the material. However, a major problem in superconductive technology is that if even a small portion of the superconductive material exceeds a critical temperature, the superconductive state is lost for that portion, and this eventually spreads over the entire material.

Currently, two methods are used to combat this problem. The first is to cover the superconductive materials with a large quantity of copper, aluminum or other metals which are both electrically and thermally excellent conductors. The second is to split the superconductive wires as finely as possible.

As shown in Figure 2-40, in practical applications these two methods are combined by embedding finely spun superconductive wires in a good conductor such as copper. The manufacture of such thin multiple core wires is one of the most important factors in creating superconductivity.

Fig. 2-38. Utilization of Superconductivity

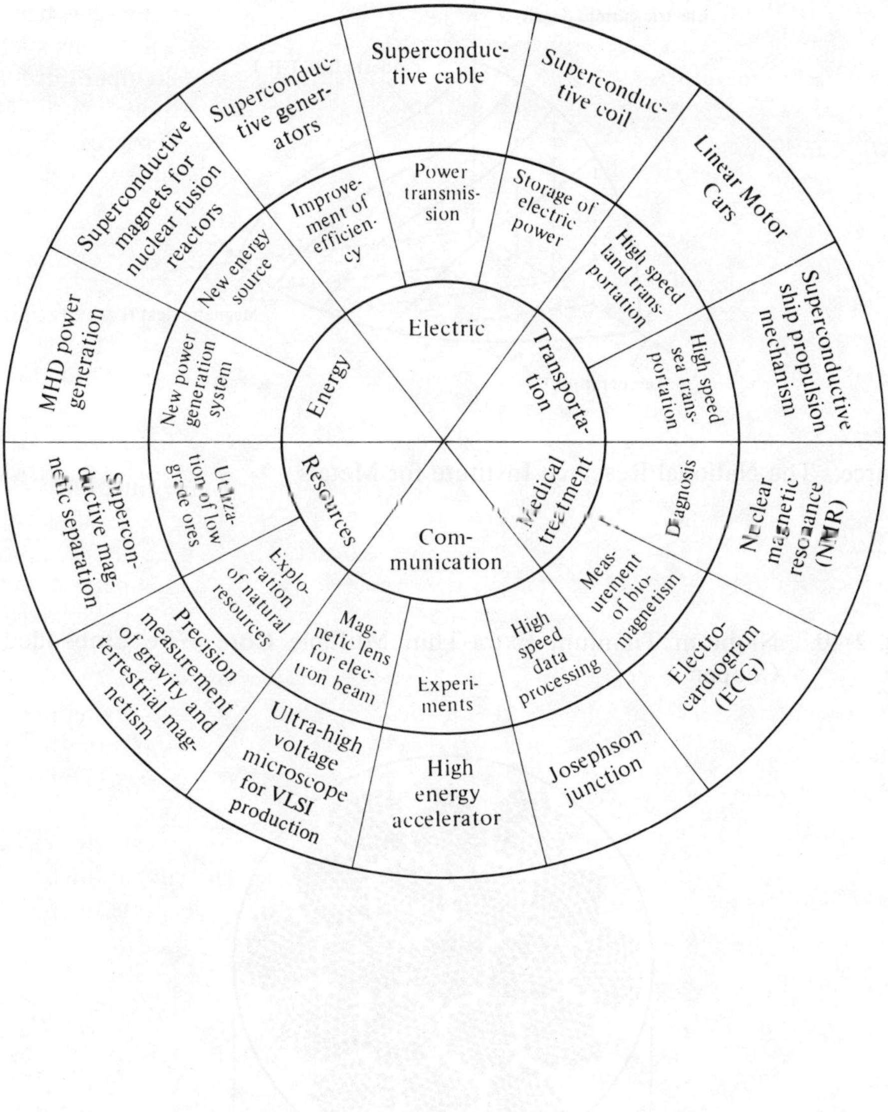

Fig. 2-39. Superconductive State

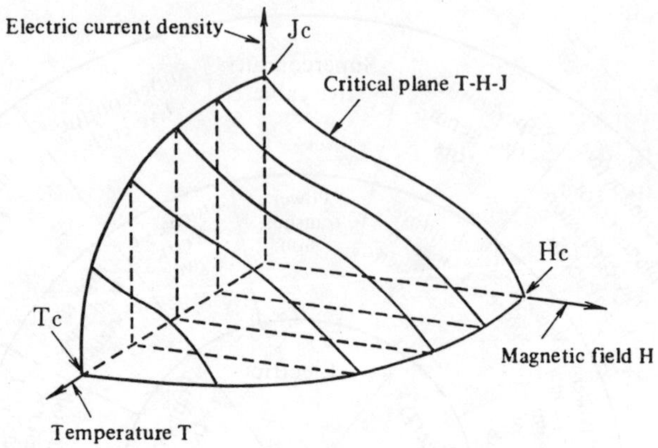

Source: The National Research Institute for Metals

Fig. 2-40. Niobium Titanium Extra-Thin Multiple Core Wire Embedded in Cooper

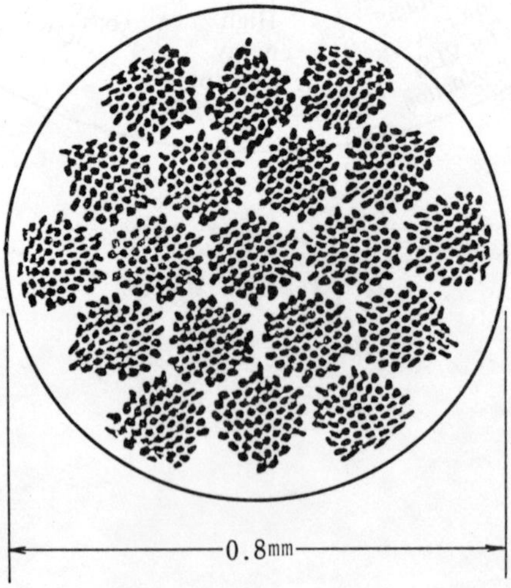

Source: National Research Institute for Metals

Figure 2-41 shows the characteristics of upper critical magnetic field and critical temperatures of the main superconductive materials at 4.2K. The black circles in the diagram indicate materials which have been put into practical application using the above-mentioned techniques, while the white circles show materials with possible future applications.

At an early stage of research, lead (PB), niobium (Nb), and other single-element metals were once categorized as superconductive materials, but, low in the upper critical magnetic field, they were later considered inappropriate for practical applications, and research on superconductivity was suspended until the discovery in the United States in the 1950's of niobium-zirconium (Nb-Zr), niobium-titanium (Nb-Ti), and other alloys. Since then, research using these materials has been accelerated in an effort to develop practical applications, and currently the alloys are widely used as superconductive materials which are both mechanically stable and easy to manufacture.

Fig. 2-41. Critical Temperatures and Upper Critical Magnetic Field of Various Superconductive Materials (4.2K)

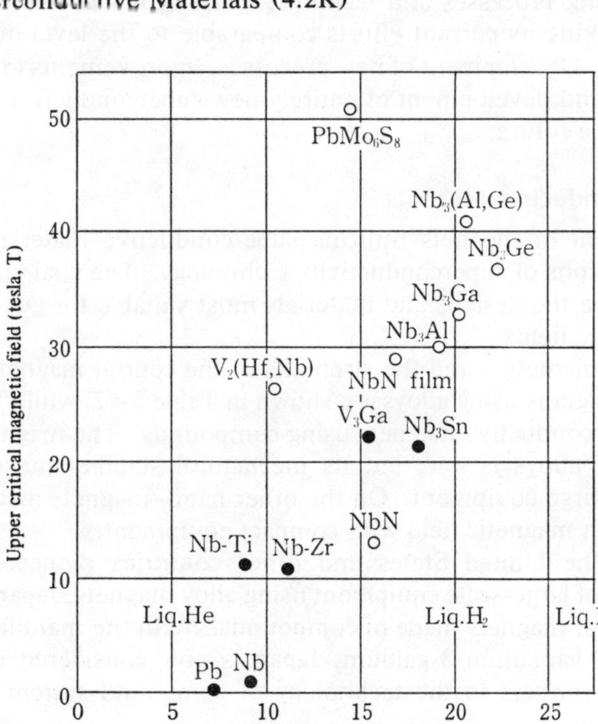

Source: National Research Institute for Metals

Superconductive materials made of niobium 3-tin (Nb_3Sn) and other metal compounds, although generally having excellent characteristics, are difficult to process into extra-fine multiple core conductors because of their brittleness. To overcome this drawback, a surface diffusion process making use of the diffusion reaction of niobium 3-tin was first developed in the United States.

To improve the practical applications of nibium 3-tin and vanadium 3-gallium (V_3Ga) as superconductive materials, a manufacturing process using the diffusion promoting effects of copper was developed in Japan. Later, a more stable composite process was developed using copper alloy. This in turn has led to the development of the *in situ* process, which permits manufacture of materials with higher performance as well as better endurance against bending and tensil stresses.

Research and development of superconductive materials was initially conducted mainly in the United States, where dramatic progress was made with the development of niobium-titanium conductors and, more recently, the new materials represented by niobium 3-tin and other highly effective materials, called type A15 compounds.

Japan has attained significant achievements in research and development of various working processes and technologies designed to improve properties of materials, making its current efforts comparable to the level in Europe and the United States. Development of new processes, improvements in the performance of materials, and development of entirely new superconductive materials remain as tasks for the future.

(b) Superconductive Magnets

Development of magnets utilizing superconductive materials is one of the major applications of superconductivity technology. The goal of current research is to determine the designs and materials most suitable for generating large and stable magnetic fields.

The inside diameters and the strengths of the central magnetic field of superconductive magnets using alloys are shown in Table 2-42, while Table 2-43 shows those of superconductive magnets using compounds. The magnetic field of magnets made of alloys is low, but its mechanical stability makes it possible to manufacture large equipment. On the other hand, magnets made of compounds produce a high magnetic field with compact equipment.

Although the United States and other countries pioneered research and development of large-scale equipment using alloy magnets, Japan has emphasized development of magnets made of compounds. With the manufacture of a strong magnet using vanadium 3-gallium, Japan is now considered to be one of the world's front-runners in the technology of compound system superconductive magnets.

Next, progress in various projects using superconductivity technology will be examined in turn.

Table 2-42. Superconductive Magnets Made of Alloys

Countries	Materials	Inside diameter	Central magnetic field
Japan	NbTiZr	0.6 x 0.95 m (Race track shape)	4.53T
United States	NbTi	4.27 m	3.0 T
Switzerland	NbTi	4.27 m	3.5 T
Japan	NbTi	2.5 x 3.5 m	6.4 T

Source: National Research Institute for Metals

Table 2-43. Superconductive Magnets Made of Compounds

Countries	Materials	Inside diameter	Central magnetic field
Unites States	Nb_3Sn	15 cm	13.5T
United States	Nb_3Sn	4 cm	15.0T
Japan	$Nb_3Sn + V_3Ga$	3 cm	17.5T
United States	Nb_3Sn	9.5 cm	14.1T

Source: National Research Institute for Metals

(c) Superconductive Magnets for Nuclear Fusion Reactors

A nuclear fusion reactor of the Tokamak type confines plasma in a magnetic field. To generate such a powerful magnetic field, a superconductive magnet is regarded as most suitable. However, since the magnet for the Tokamak reactor must be large and sophisticated, with a huge refrigeration and cooling system, it is different from superconductive magnets used for other purposes.

Research and development of superconductive toroidal magnets is now being carried out under multi-national cooperation between the United States, the EC, Switzerland, and Japan, with a target magnetic field strength of 8 tesla and 130 megajule. In 1982, performance tests on a magnet developed by the Japan Atomic Energy Research Institute achieved the target specifications for the first time in the world.

Magnets of similar performance are scheduled to be completed in the United States around 1983 and in the EC and Switzerland around the end of 1983. Superconductive toroidal magnets that can pulsatively generate an intensive magnetic field for heating and control of plasma are also needed. A magnet of this type, with a magnetic field of 3.3 megajules, has been built in the United States, and one with a similar capacity is also under development in Japan by the Electrotechnical Laboratory of the Ministry of International Trade and Industry.

Research and development of nuclear fusion reactors in Japan was started at the same time as in Europe and the United States, and reactor core technology in Japan is roughly comparable with those countries.

In order for a nuclear fusion reactor to be economical, large-scale magnets are required. Therefore, further research must be directed toward production of specially designed large-scale magnets, including development of materials which are more stable and capable of creating a strong magnetic field.

(d) Superconductive Power Generators

The basic design of a superconductive power generator is almost the same as that of a conventional power generator, the difference being that a superconductive magnet is used for the magnetic coil of the rotor. This makes it possible to reduce energy loss to half or one-third of conventional ones, producing a generator both high in efficiency and light in weight.

As illustrated in Figure 2-44, an experimental small-scale superconductive generator was manufactured for the first time in the United States in the early 1970's. Since then, research and development of large-scale generators has been increasingly promoted in various countries. In Japan, 30,000kW and 50,000kW power generators have been successfully developed by private enterprises, and the world's largest generator, completed in 1982, is now being test operated.

Fig. 2-44. World Production of Prototype Superconductive Power Generators

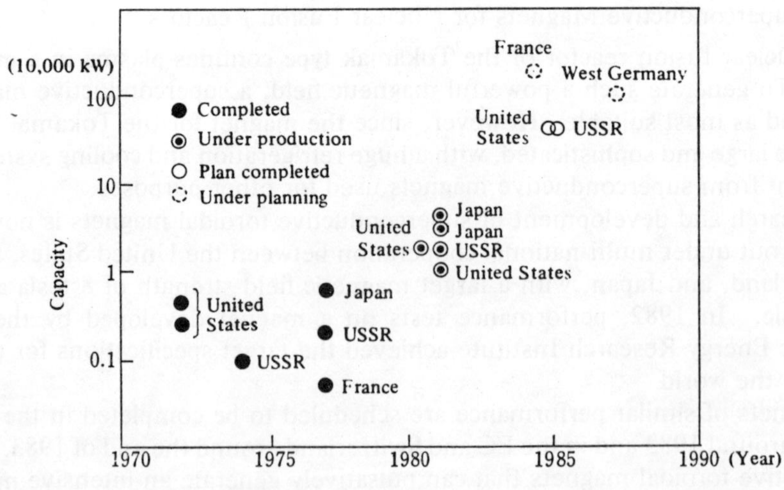

Source: The Japan Science Foundation

Superconductive power generators are also under development in the United States, France, West Germany, the Soviet Union, and other countries. Power generators larger than those produced in Japan are now being developed in the United States and the Soviet Union, both of which are planning to make practical applications around 1990.

Although its level of technology is as advanced as that of other countries, Japan, still without any projects on as large a scale as the United States and the Soviet Union, is expected to actively promote overall research and development for practical applications.

(e) Linear Motor Cars

Linear motor cars, using superconductive magnets, travel and levitate due to magnetic force. Because no contact is made with the rail, they are significantly energy efficient as well as low in noise and vibration, and can reach speeds of over 500km per hour, much higher than conventional trains.

Currently, Japan, the United States, West Germany, and other countries are developing linear motor cars. In Japan, research was initiated in 1962 by the Japanese National Railways (JNR), and an experimental train (the ML500) was built and tested in 1979. This train, levitating 10cm, reached a maximum speed of 517km per hour, the world's highest. In 1982, the JNR succeeded in testing the

Fig. 2-45. Cross-section of the MLU001

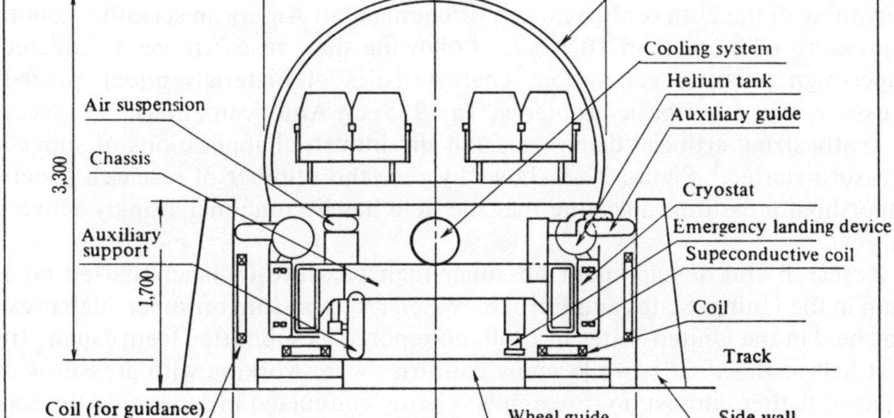

Source: Japanese National Railways

world's first manned experimental train, the MLU001, which reached a maximum speed of 262km per hour. A cross-section of the MLU001 is shown in Figure 2-45.

The superconductive magnets in linear motor cars require a compact and highly reliable refrigeration and cooling system. For such trains to be put into practical use, a great deal of research in this and other systems remains to be done.

As outlined in the above sections, to effectively utilize superconductivity created under extremely low temperatures, progress is needed in development of a reliable refrigeration system, effective conductive materials, structural design of magnets, and instrumentation and control technology, to mention just a few fields. Japan, realizing the importance of these fields, is developing technology for synthesizing and processing superconductive materials, methods for analyzing characteristics of materials at extremely low temperatures, highly efficient fluid helium cooling systems, and extremely low temperature, energy-saving projects.

(2) Super-high Pressure

Super-high pressure of several hundred to several thousand bars (1 bar ≑ 1 atm) is used in the synthesis of ammonia, methanol, polyethylene and other valuable substances for industrial use.

Industrial-use diamonds are also made under high pressure of more than 10 kilobars. Research on the physical properties of inorganic substances, high-speed molding of metals and pressure-welding of metals under super-high pressure are also being investigated, all of which show promise of leading to the development of technology of great importance to a country such as Japan, highly industrialized but lacking natural resources.

The use of super-high pressure in modern technology was established at the beginning of the 20th century, when Bridgeman, an American scientist, generated a pressure of more than 10 Kbar. Following that, research was conducted on super-high pressure generation, characteristics of materials under super-high pressure, and other basic problems. In 1955, an American enterprise succeeded in synthesizing artificial diamonds, and the industrial applications of super-high pressure started. Figure 2-46 shows by year the number of research papers on super-high pressure, indicating that the field has become increasingly active year by year.

Research and development on super-high pressure in Japan started far later than in the United States and Europe. When a symposium on super-high pressure was held in the United States in 1960, no report was submitted from Japan. In the first half of the 1960's, while many countries were working with pressures of 20 Kbar or higher, almost no research was being conducted in Japan. By the second half of the 1960's, however, domestic research on super-high pressure finally grew active, and an international conference on high pressure technology was held in Japan in 1974, signifying the beginning of Japanese activity in the field at an international level.

Fig. 2-46. Annual Number of Research Papers on Super-High Pressures (10 Kbar or higher)

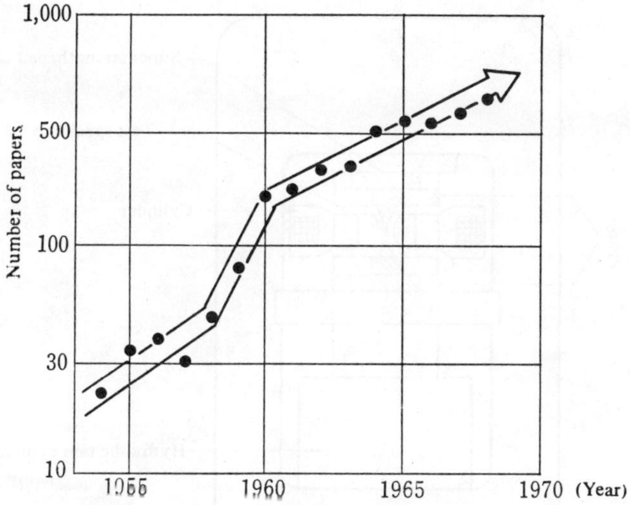

Source: The Science and Technology Agency

Although pressure control and measurement technology will be important for practical applications of super-high pressure technology, the most fundamental problem is the actual generation of super-high pressure. This is achieved either through a static method using piston cylinders, or through a dynamic method using explosives, each of which is described below.

(a) The Static Generation Method

The static method generates pressure in a container through compression, as shown in the model in Figure 2-47.

To synthesize materials under super-high pressure, a large container increases the efficiency of the process. However, in practice the size of container is sacrificed to generate higher pressure. The key point of current research and development is the generation of higher pressures in large containers.

Figure 2-48 shows pressures and volumes attained with the static method. In 1981, the Soviet Union attained the highest pressure in the world, and in the same year, the National Institute for Research in Inorganic Materials attained the highest value in Japan. The development of a device capable of exceeding the Soviet record is now at the planning stage in Japan, where research in this field has made outstanding progress in recent years.

Fig. 2-47. Model of a Static Super-high Pressure Generating Device

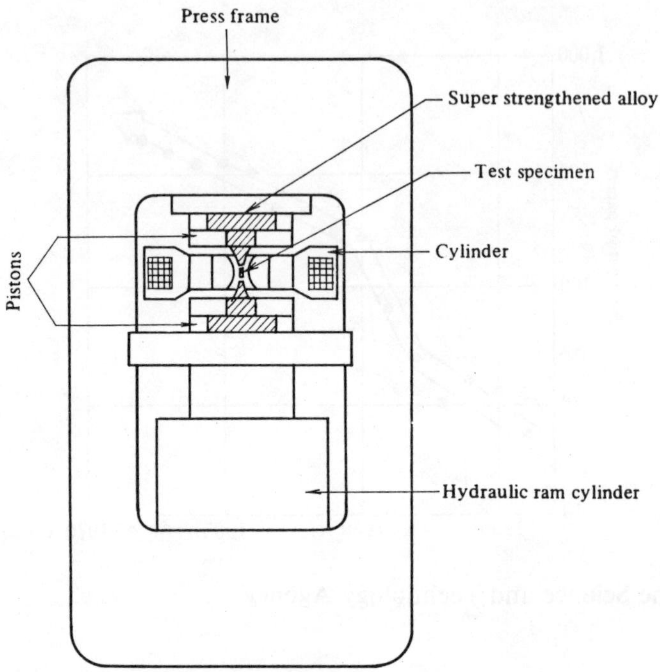

Source: The Science and Technology Agency

(b) The Dynamic Generation Method

When substances collide at very high speed, the resultant shock wave creates high compression for an extremely short time, until the wave passes away. The dynamic generation method utilizes this principle to create super-high pressure, using technology and knowledge of high-speed collisions to control the shock waves. An important feature of this method is that it generates higher pressure and higher temperature and does so more easily than the static method, but only for a very short time.

The first device developed for generating shock waves is the explosion lens shown in Figure 2-49. Research using this lens was initiated by the United States and the Soviet Union and later followed up by European countries. Subsequently, a large-scale gas accelerator was developed for conducting more precise shock wave experiments. The two-stage gas acceleration device shown in Figure 2-50 has achieved remarkable success. The device, developed in the United States in 1948, was first used for experimental generation of super-high pressures in 1966. At present, using this and similar devices, it has become possible to generate pressures of several Mbar.

Fig. 2-48. Pressures and Volumes Attained with Static-Method Devices

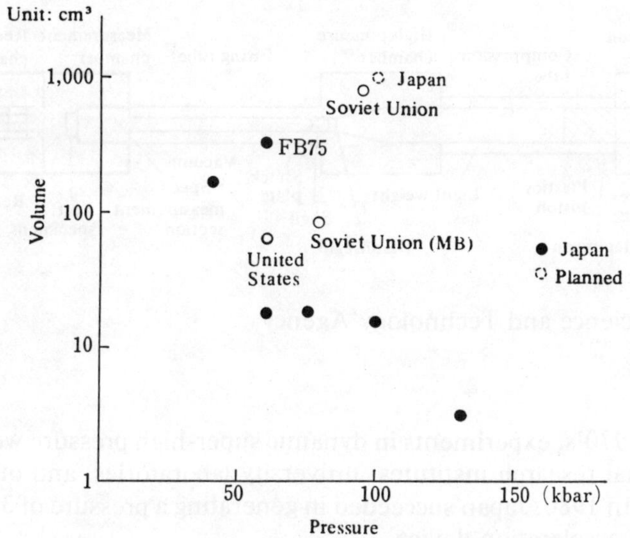

Source: The Science and Technology Agency

Fig. 2-49 Model of an Explosion Lens

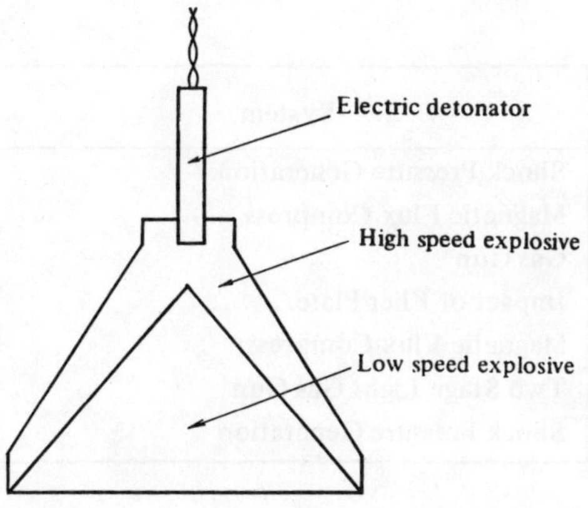

Source: The Science and Technology Agency

Fig. 2-50 Two-Stage Light-Weight Gas Acceleration Device

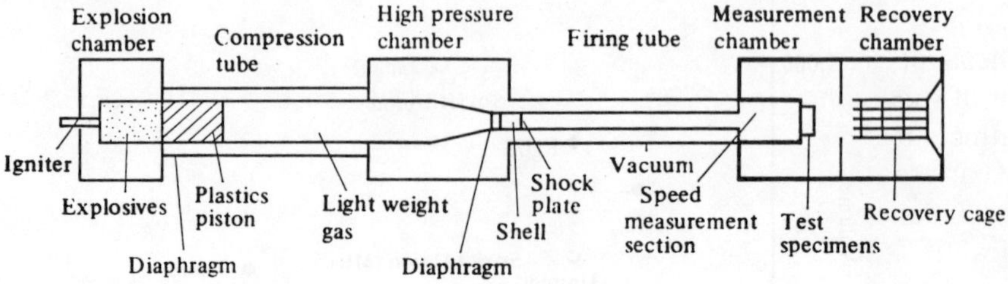

Source: The Science and Technology Agency

In the early 1970's, experiments in dynamic super-high pressure were started in Japan at national research institutes, university laboratories, and other research organizations. In 1980, Japan succeeded in generating a pressure of 3.6 Mbar with a two-stage gas acceleration device.

Table 2-51 shows accomplishments attained with various dynamic methods by the United States, the Soviet Union and Japan. Although Japan is still behind the United States and the Soviet Union, research in generation of pressure, measurement methods, and other critical problems is being actively promoted.

Table 2-51. Present State of Technology of the Dynamic Generation Method

Countries	System	Generated pressures
United States	Shock Pressure Generation	6.4 Mbar
United States	Magnetic Flux Compress	10 Mbar
United States	Gas Gun	7 Mbar
Soviet Union	Impact of Flier Plate	10 Mbar
Soviet Union	Magnetic Flux Compress	25 Mbar
Japan	Two Stage Light Gas Gun	2~4 Mbar
Japan	Shock Pressure Generation	3~4 Mbar

Source: The Science and Technology Agency

Although super-high pressure technology offers an extremely wide range of applications, including synthesis of new materials, not many years has elapsed since full-scale research was initiated, so that the field is still at the basic stage. The objectives of technology using static super-high pressure generation are the synthesis of artificial diamonds under pressure of about 100 Kbar and the synthesis of superconductive materials under pressure of several Mbar. In the field of dynamic super-high pressure technology, methods of generating pressures over 20 Mbar are needed to further boost the engineering of new materials.

The Japanese government intends to emphasize (1) development of practically applicable large-scale equipment using the dynamic method to generate super-high pressures of 10,000 Kbar; (2) development of large-scale equipment using the static method to generate super-high pressures of 100 Kbar in a volume of 1,000cm^3; and (3) research on practically applicable material synthesis technology.

New Materials

Materials have played an important role in all technical innovation, from the simple hand tools of the ancient Iron and Bronze Ages to the electronics technology now revolutionizing modern civilization.

Interest in new materials has greatly increased recently, partly because of developments in a variety of pioneering fields such as space technology and nuclear power, which depend greatly on new materials and partly as a result of growing demands for energy efficient materials.

These new materials include (1) fine ceramics (also called new ceramics) which offer a wide range of applications; (2) new polymer materials, including highly functional separation membranes and conductive plastics; (3) shape-memory alloys, amorphous alloys and other new metals; and (4) composite materials such as carbon-fired reinforced plastics (CFRP). These and other new materials can be utilized in a wide range of industries, not only as structural but also as functional materials with various characteristics.

Because new materials lead to technological innovations in various fields, it is essential for Japan to promote research and development in this area in order to reach the European and American level in the future. With this in mind, new materials technology was adopted as one of the special projects in the fiscal 1982 budget for promotion of science and technology, as a result of which, research on high performance materials has been initiated. In addition, research on the creation of new materials utilizing a zero gravity environment was started to explore the possibility of manufacturing new materials aboard the U.S. space shuttle.

(1) Fine Ceramics

Ceramics are made of non-metallic inorganic materials, which have been fired under high temperature. New ceramics have been created from selected refined materials of a certain chemical composition, fired under controlled conditions to bring out their best characteristics.

Depending on the materials used and the manufacturing process, fine ceramics demonstrate a number of excellent properties, such as heat resistance, durability, electromagnetism, and photo-sensitivity. They can be found in a wide range of applications, as illustrated in Figure 2-52. As materials that can withstand high temperatures, high stress, radiation and other severe conditions, new ceramics are expected to be especially useful in such fields as space, marine, and nuclear power development.

Because raw materials for new ceramics are abundantly available and create a high added value in finished products, as industrial materials they are considered to have a very promising future.

Fig. 2-52. Applications of New Ceramics

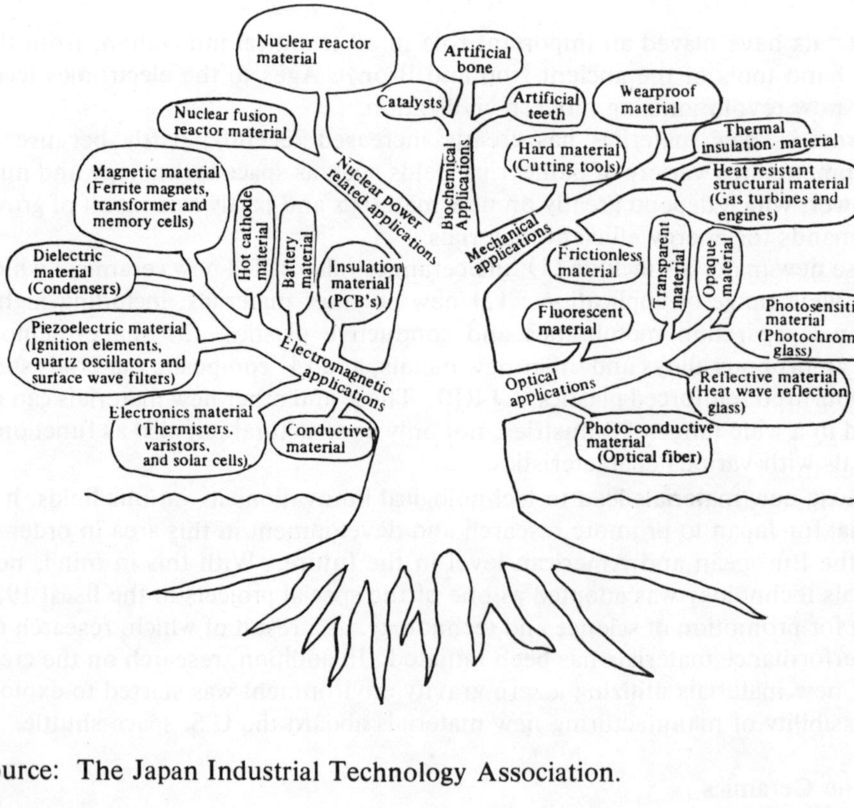

Source: The Japan Industrial Technology Association.

Among the various kinds of fine ceramics, those used in electronics have advanced most extensively, and are widely used in integrated circuits, various sensors, and many other devices. The technological level in this field is advanced in Japan, as witnessed by the high international demand for Japanese-made IC packages. In the medical field, artificial bones and teeth made of new ceramics have also reached the stage of practical applications.

Since fine ceramics are high in heat resistance and enhance the efficiency of engines, in recent years, as shown in Table 2-53, they have attracted world attention as materials for engines and gas turbines. Research and development in this field has been promoted mainly in the United States and West Germany, beginning in the early 1970's. Although full scale research and development in Japan started later, Japanese technology has developed ceramics with the highest resistance to bending, and has produced the world's first successful ceramic diesel-engine car.

For use in engines, ceramics made from silicon nitride and silicon carbide are considered most suitable, because of their ability to resist heat, water, and corrosion. However, they are brittle and difficult to sinter, so that many problems still remain to be solved, such as manufacture of finely powdered materials of high purity, forming and sintering techniques for obtaining high-precision ceramics, proper assessment technique, and flow defection.

Development of new ceramics has since 1978 been one of the major targets for the creation of an efficient gas turbine engine under Japan's Moonlight Project, concerned with energy-saving technologies.

(2) Polymers

Since the 1970's, technological innovation in the field of polymer materials, represented by plastics, has been advancing rapidly. One example is the trend of replacing metals with light, corrosion-proof plastics, which have improved heat resistance and strength. In recent years, new polymer materials, called engineering plastics, have been developed, replacing metallic parts in cars and cameras. Another trend can be seen in the development of functional polymer materials, which utilize various properties of their molecular structures, the major applications of which are shown in Table 2-54.

Table 2-53. Research and Development of Ceramic Engines

Country	Organization		Outline
Japan	Public laboratories		Super heat resistant ceramics for a highly efficient gas turbine engine is being studied under the Moonlight Projects. Also ceramics for industrial structural materials with high heat resistance, corrosion-proof, precision, and reliability is being developed.
	Private companies	A	Started development of ceramic engines in 1965, and developed ceramics with high resistance to bending. Currently engaged in establishing tests to measure the service life of ceramic parts
		B	Developing a gas turbine engine for automobile with an inlet temperature of 1,350°C. Conducted simulation tests in 1980 using combustor and rotor made of silicon nitrate and silicon carbide.
		C	Succeeded in purifying silicon nitride powder which can be made into high bending resistant material through a joint research with another private company. Manufactured large ceramic parts with three-point bending strength of $80 Kg/mm^2$.
		D	An application research on diesel engine with ceramic parts, conducted jointly with another private company, was selected as one of the projects subsidized by the MITI. Succeeded in a test run of an automobile with a 3-cylinder, 2,800 cc ceramic diesel engine in 1981.
		E	Succeeded in running a 2-cylinder, 50 cc ceramic engine made of silicon nitride at 3,000rpm for 50 hours.
United States	Public laboratories		Development of ceramic parts for turbine engines was started in 1971, and ceramic engine in 1979 under the Dept. of Energy.
	Private companies	I	Developing an automobile gas turbine engine with durability of 200 hours.
		J	Adopted silicon nitride ceramics for gas-fire nozzle, shroud, and heat exchanger of a gas turbine engine.
		K	Developed a gas turbine engine with stator, rotor, and combustor made of silicon nitride ceramics.
		M	Succeeded in an operation test of a single cylinder diesel engine made of silicon nitride for 250 hours.
West Germany	Public laboratories		Development of ceramic parts for gas turbine engines was started in 1974
	Private companies	N	Developing 350 horsepower gas turbine engine for trucks
		P	Developing gas turbine engines for automobile
		Q	Developed 175 horsepower gas turbine engine and performed tests at an inlet temperature of 1,370°C. Announced a car with gas turbine engine using a ceramic rotor.

Source: The Science and Technology Agency

Table 2-54. Applications of Functional Polymer Materials

Functions	Major applications
Ion exchange	Production of caustic soda, waste water treatment, and recovery of uranium from sea water
High polymer separation membrane	Desalination of sea water, artificial kidney, enrichment of oxygen, and separation of water and alcohol (alternative for distillation)
Electrical conductivity	Photo-sensitive recording, solar cells, sensors, superconductors, light-weight electric conducting materials
Piezoelectric and pyroelectric properties	Sensor, transducer, and sound system
Photo-sensitivity	Photoresist for IC's and plating for printing
Biomedical application	Artificial blood and organs

Source: The Science and Technology Agency

The separation of substances in a liquid or gas mixture is one of the most promising applications. Polymer materials have already been given practical applications in ultrafiltration of low-weight molecules, reverse osmosis for extracting a solvent from a solution, dialysis for extracting solutes, and ion exchange. Japan is one step behind the United States in the development of the reverse osmosis process, but has recently succeeded in developing such a process for sea water desalination which has the world's highest desalting ratio and cost efficiency. In addition, the Japanese level of research and development of artificial kidneys is the highest in the world.

Use of polymer materials to separate substances requires a far smaller amount of energy than conventional methods and also makes it possible to separate substances with similar boiling points or of the same composition. One example is an oxygen enrichment membrane, which by separating oxygen from normal air can raise the oxygen concentration to around 40 percent. Now approaching the stage of practical application, this will be useful for persons with respiratory problems, and can also be utilized to increase the efficiency of engine combustion systems.

Use of polymer materials to recover uranium from sea water is also attracting world attention. Although large amounts of uranium are believed to exist in sea water, its concentration is extremely low. Of the various methods so far tried, the adsorption process is presently considered best. In experiments carried out in England in the 1960's, hydrous titanium oxide, an inorganic substance, was believed to be the best adsorption substance.

Recently, however, a polymer material called chelating resin, which selectively adsorbs uranium, has also shown much promise, and though previously used for recovery of mercury and other heavy metals from industrial waste water, it was used for the first time in Japan for the recovery of uranium from sea water. Research is being carried out at the Government Industrial Research Institute in Shikoku, private enterprises, universities, and other research laboratories. In other countries, most research has focused on use of hydrous titanium oxide as an adsorbent, but research on the use of polymer materials was also recently started in the United States and West Germany. The amidoxime adsorbent studied in Japan is a kind of chelating resin with excellent adsorption capacity and speed, as shown in Figure 2-55, but further improvement of its durability and its adsorption speed per unit volume is necessary.

These new polymer materials have excellent prospects as energy-efficient substitutes for conventional materials and as new industrial and medical materials.

As objects of research and development in Japan, fine polymers, high efficiency separation membrane polymers, electro-conductive polymers, and high crystalline polymers are being emphasized for their future industrial applications.

Fig. 2-55. Adsorption of Uranium from Natural Sea Water

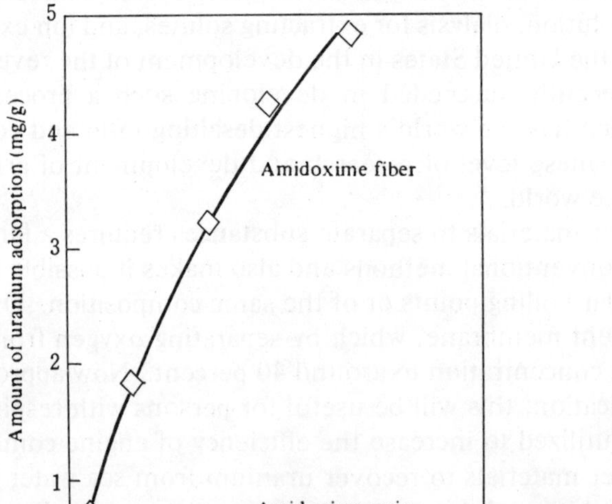

Source: The Agency of Industrial Science and Technology

(3) New Metals

The majority of metals are used as structual materials, and today's industrial society is extremely dependent on various kinds of metals. With the increasingly diversified applications of metals in recent years, entirely new functions have come into demand. Consequently, other than the improvement of existing metals, new directions in metal research have appeared.

One trend is the development of new metal alloys with new functions, typical examples of which are the shape memory alloy and the hydrogen occlusion alloy.

When deformation exceeds a dynamic limit of around 0.5 pecent, conventional metals cannot return to their original shapes. However, certain kinds of alloys memorize the shape set at high temperatures, and even after deformation resume this shape when reheated. This is called a shape memory effect, and an alloy with this characteristics, consisting of nickel and titanium, was invented for the first time in the early 1960's in the United States.

Since then, shape memory alloys have been applied to joints, automatic switching devices used in greenhouse doors, and various other applications. Recently, however, such alloys have been applied for medical treatment, space-craft antennas, thermal engines, and in other fields. A thermal engine utilizing the shape memory alloy, although not so efficient, works on subtle temperature differences, and is considered most effective in recovering energy from solar heat and industrial waste heat. In Japan, a nickel-titanium alloy with excellent resistance to wear and corrosion was developed in the early 1960's. But in recent years it has been studied as a shape memory alloy. Today, the technology of shape memory alloys in Japan has reached the world's highest level, in parallel with that of the United States. Copper alloys with a shape memory effect, although less efficient than nickel-titanium alloys, are cheaper and are being studied mainly in Europe.

Although it has long been known that metal occuludes hydrogen, recently research on use of metal hydrates to store energy has been actively carried out. To transfer or store hydrogen, high pressure cylinders are normally used, but use of a hydrogen occlusion alloy improves safety and also offers advantages in capacity and weight, as shown in Table 2-56. Research on hydrogen occlusion alloys is active in the Netherlands and the United States, and a lanthanum and nickel alloy that discharges hydrogen occluded at atmospheric temperature was achieved in 1970.

In Japan, as part of the Sunshine Project started in 1974, research is being carried out on use of hydrogen occlusion alloys to transfer and store hydrogen. Moreover, development of a hydrogen occlusion alloy as a means of storing heat energy is also being promoted.

Table 2-56. Comparison of Hydrogen Transfer Methods

Transporting methods	Volume (ℓ)	Weight (kg)	Amount of hydrogen (Nm^3)
Conventional method (gas cylinder of 47 ℓ and 150 atm)	47	55	7
Hydrogen occlusion alloy (Mg–5%Ni cartridge)	10	13	7

Source: "Sunshine Project News," publication of the Agency of Industrial Science and Technology of the Ministry of International Trade and Industry (MITI)

Another trend in recent research is aimed at adding new characteristics to metals, exemplified by ultra-fine particle metals and amorphous metals.

When a metal is made into ultra-fine particles of from 10 to 2,000A (1A, or angstrom, = one hundred millionth of a cm) in diameter, it assumes peculiar physical properties, as shown in Table 2-57. This discovery was announced in 1962 in Japan, where process technology for it was also later developed, making the country the world leader in this field.

On the other hand, research on applications is being conducted mainly in Europe and the United States. The United States and France produced an experimental, extremely low temperature dilution refrigerator using ultra-fine particle silver developed in Japan, and recorded a temperature of 0.02K to 0.002K. Though there are still many uncertain factors, ultra-fine particle metals are expected to lead to a wide range of applications in the future, so that several research projects are being promoted in Japan to study physical properties, applications, and process techniques.

Amorphous (noncrystalized) metal has a number of properties different from those of conventional metals, including corrosion resistance and soft magnetism. Amorphous metal is made by cooling molten metal very rapidly. Ever since the first amorphous metal was made from an alloy of gold and silicone in the United States in 1960, a number of other amorphous metals have been discovered. Since the invention in 1970 of the liquid quenching process for making amorphous metal in wire rod or plate form, interest in applied research has soared. Examples of amorphous metal applications are shown in Table 2-58.

Research and development in the United States is promoted mainly at the research laboratories of private enterprises, with a number of universities also taking part in various projects. Since 1978, about $6 million has been invested in the development of energy efficient amorphous magnetic materials in the United States.

Table 2-57. Outstanding Properties Demonstrated by Ultra-fine Particle Metals

- Magnetic property . . . . . . A certain kind of iron alloy demonstrates far more intensive magnetic property than other heavy substances
- Water-repellent property . . High surface activity can be obtained
- Combustibility . . . . More heat is generated when an extremely small amount of ultra-fine particles is added in the rocket fuel
- Catalyst . . . . The surface area is vast (70 m^2 per 1 g)
- Low temperature sintering property . . . . . . The melting point is extremely low in contrast with other heavy substances and the sintering temperature is low (The ultra-fine particles of mercury can be sintered at 100°C or lower, and that of nickel at 360°C or lower)
- Heat exchanging property . . . . . . There is almost no thermal resistance at low temperature
- Photo-absorption characteristics . . . . The chromium alloy has good photo absorption characteristics
- Superconductivity . . . . . . Some ultra-fine particles have crystalline structures suitable for superconductivity
- Large in surface tension, and the internal pressure is equivalent to that of the earth (several 100,000 atm).

Source: The Research Development Corporation of Japan

Table 2-58. Applications of Amorphous Metal

Function	Main application
High strength	Composite materials and springs
Corrosion resistance	Cutting tools, plate and wire materials
High elasticity	Springs for precision equipments
Magnetic property	Magnetic recording head and low loss transformer
High electric resistance	Strain gauges and resistors
Catalyst	Chemical reaction and electrolysis of caustic soda

Source: The Research Development Corporation of Japan.

In Japan, on the other hand, research initially carried out at a metal research institute of Tohoku University was in 1977 expanded and adopted as a subject for applied technology at the Research Development Corporation, where, after three and half years of research, industrialization of amorphous metal technology is near completion. During this period, research and development at other universities, research laboratories, and private enterprises also became active. Amorphous metal has also been selected as an area of research under the System for Exploratory Research for Advanced Technology.

As described above, amorphous metal is very simple to manufacture and offers significant energy savings in addition to its other advantages. As a result, research and development in this field has become active worldwide, led by Japan and the United States, where full-scale research and development started almost simultaneously.

(4) Composite Materials

Composite materials consist of a number of different materials which when combined create new characteristics not found in the original separate materials. Composite materials, unlike single-element materials, also allow an optional distribution of characteristics. Although composite materials come in a variety of combinations depending on the component materials, research has been mainly carried out in the fields of fiberglass reinforced plastics (FRP) and fiberglass reinforced metals (FRM).

Among FRP's, carbon fiberglass reinforced plastic (CFRP) is an excellent composite material, with the strength of metal as well as the lightness and easy molding qualities of plastics. For this reason, CFRP's are used throughout industries ranging from aerospace to fishing rods and golf clubs. In airplanes, their outstanding weight reduction qualities have resulted in improved energy savings; for example, CFRP and other composite materials amounted to about 1.9 percent, or 1.5 tons, of the total weight of the Boeing 767, and about 2.6 percent of the Boeing 757.

The carbon fiber mainly used in CFRP's is either of the pitch or the polyacrylonitrile (PAN) type. Although the cost of the PAN type is about 10 times that of the pitch type, it has excellent characteristics such as strength and elasticity and is therefore highly valued as a composite material. The PAN type carbon fiber was developed at the Government Industrial Research Institute in Osaka in 1959, and demand for it has recently expanded rapidly, such that about 70 percent of the world's 1980 total demand of 815 tons was supplied by Japan. Japan has the world's highest level of technology in carbon fiber as a CFRP material, while Europe and United States have higher levels of application techniques gained through aerospace research and development.

Full-scale research on FRM's started in the 1970's. Though a relatively new composite material with many problems yet to be solved, because its rigidity and heat resistance are greater than CFRP's, it is regarded as a promising material in the aerospace industry.

Fig. 2-59. Breakdown of World Demand for PAN Carbon Fiber (1980)

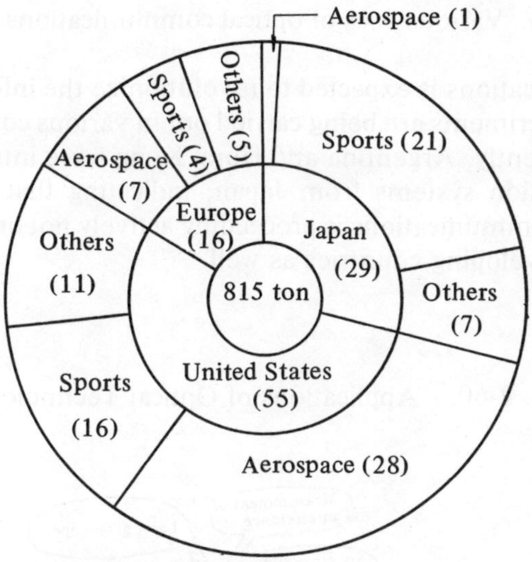

Source: A carbon fiber manufacturer

As seen in Figure 2-59, carbon fiber in Japan is little used in applications requiring very high standards, such as in the aerospace industry. Consequently, even though Japan has first class technology for production of materials, it is behind Europe and the United States in research in more durable materials such as FRM's.

Electronics and Communications

Electronics and communications technology have played an important role in improving standards of living and developing society and economy. Among the variety of such technologies, trends in optical technology and devices such as semiconductors will be discussed here.

(1) Optical Technology

Optical technology had been mainly limited to telescopes, microscopes, cameras, and other optical devices until invention of the laser in 1960. Since then, optical technology has progressed rapidly and is now being used in such areas as material processing, medical treatment, and telecommunications, as shown in Figure 2-60.

Among these various applications, the one which is attracting most attention is communications. Optical communication technology, using small, light-weight optical fiber as a transmission medium, makes it possible to transmit large quantities of information in a wide band with little transmission loss and no induction disturbance. Furthermore, quartz, the main raw material of optical fiber, exists abundantly on earth. Various uses of optical communications are shown in Table 2-61.

Optical communications is expected to revolutionize the information society of the future, and experiments are being carried out in various countries as indicated in Table 2-62. Recently, Argentina and Hong Kong have introduced large-scale optical communication systems from Japan, indicating that the use of optical systems in public communications is proceeding actively not only in industrialized countries but in developing countries as well.

Fig. 2-60. Applications of Optical Technology

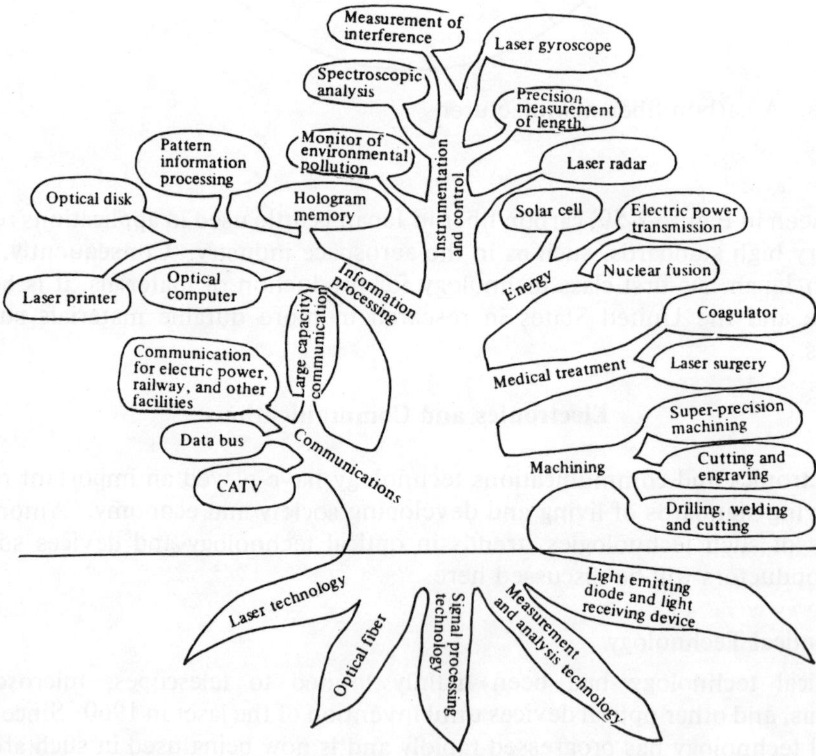

Source: The Science and Technology Agency

Table 2-61. Uses of Optical Communications

Field	Major application	Major advantages
Telephone network	1. Toll circuit 2. Inter-office circuit 3. Subscriber's line 4. Submarine transmission	1. Wide band characteristics 2. Enlargement of relaying interval 3. Reduction of cross section 4. Saving of resources
Wide band network	1. CATV 2. Wired city (community) 3. Public communication for education, medical treatment, and others	1. Enlargement of relaying interval 2. Wide band and flat transmission 3. Free from interference
Data bus	1. Computer Connection with decentralized systems Inter-system connection	1. Free from interference 2. Reduction of cross section 3. Countermeasure against potential different at contact points
	2. Process control	Free from interference
Aircrafts and ships	Internal communications	Light weight
Electric power and railways	1. Operation of electric power system 2. Railway communications	1. Free from induction disturbance 2. Enlargement of relaying interval

Source: The Electronics and Communications Society of Japan

Table 2-62. Experiments on Optical Fiber Communications (1976 to 1979)

Application	Japan		United States and Canada	Europe and Australia
	Small-scale	Large-scale		
Public communications	6		15	7
Electric power and railway services	15	5	7	2
Broadcasting and CATV	14	15	10	5
Local Area Network	8	20	5	2
Military			12	3
Total	43	40	49	19
	151			

Source: The Japan Techno-Economics Society

In Japan, optical communications systems were put into use at a comparatively early date by electric power companies because of their freedom from induction disturbance due to lightning, high voltage lines, etc. For applications in public communications, tests are being carried out by Nippon Telegraph and Telephone Public Corporation (NTT) for the purpose of completing by 1984 a trunk communication system running from one end of the country to the other. Because it is possible to substantially lengthen the distance between relays and to transmit large quantities of information, research on optical cables to replace conventional submarine cables has also been carried out actively by NTT and the Kokusai Denshin Denwa Co., Ltd. (KDD).

By the early half of 1981, the number of optical communications systems installed in Japan exceeded 300, in a variety of applications as in Figure 2-63.

Fig. 2-63. Use of Optical Communication Systems by Industry

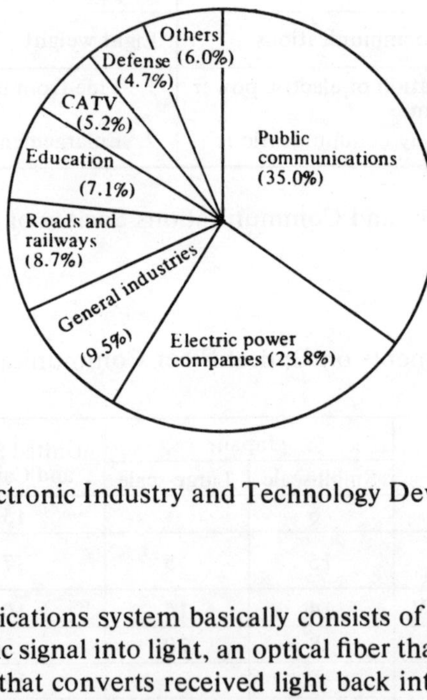

Source: The Opto-electronic Industry and Technology Development Association

An optical communications system basically consists of a light-emitting device that converts an electric signal into light, an optical fiber that transmits light, and a light-receiving device that converts received light back into an electric signal.

The optical fiber must transmit a light signal stably with small loss. As shown in Table 2-64, optical fiber has been used in the medical field since the 1950's. At that time, it was not considered possible to use optical fibers for communications because transmission loss was very large. A low-loss optical fiber system feasible for communications was attempted in the 1960's, and research started in various countries, mainly the United States.

In 1972, a glass manufacturer in the United States succeeded in making an optical fiber with a loss of 20 dB/km (at one km, transmission of 10% of light) by the CVD process (chemical vapor-phase deposition process). This breakthrough led to rapid progress in various countries.

Although research in Japan started later than in the United States, NTT in 1976 developed a fiber with the world's lowest transmission loss, as shown in Figure 2-65. In 1979, moreover, an optical fiber with an extremely low loss of 0.2 dB/km (97.7% transmission rate), almost equivalent to the theoretical limit of quartz fiber, was achieved, putting an end to competition in quartz fiber development.

Table 2-64. Developments in Optical Fibers and Semiconductor Lasers

Year	Optical fiber	Semiconductor laser
	(Optical fiber for medical use was invented in 1951)	
1957		• Semiconductor laser proposed (Japan)
1960		• Oscillation of laser for the first time (United States)
62		• Oscillation of GaAs semiconductor laser (United States)
64	• Graded index optical fiber proposed (Japan)	
1970	• Development of low loss optical fiber (United States) [20 dB/km]	• Continuous oscillation of GaAlAs semiconductor laser at room temperature (Japan, United States and USSR)
74	• Invention of low loss optical fiber manufacturing process (MCVC process) (United States) [1 dB/km]	
76	• Development of extra-low loss optical fiber (Japan) [0.5 kB/km]	• Continuous oscillation of GaInAsP semiconductor laser in 1.3 μm band at room temperature (Japan and United States)
77	• Development of a new continuous optical fiber manufacturing process (Japan)	
79	• Development of extra-low loss optical fiber (Japan) [0.2 dB/km]	• Continuous oscillation of GaInAsP semiconductor laser in 1.5 μm band at room temperature (Japan and United States)
1980	• Development of extra-low loss wide band optical fiber (Japan)	

Source: The Science and Technology Agency

NTT also developed a new manufacturing process for optical fiber called the VAD process (vapor-phase axial deposition process), with outstanding economic advantages and high productivity.

Although for quartz fiber the transmission loss has been reduced nearly to the theoretical limit, use of fluoride glass and other new materials is expected to make it possible to reduce the loss to less than 0.01 dB/km (99.9% transmission rate). Such ultra-low loss fiber is especially important for submarine cables, primarily because the distance between relays can be longer.

Light-emitting devices for optical communications such as light emitting diodes (LED) and semiconductor lasers must convert electric signals into optical signals faithfully and efficiently, in a wave length allowing the lowest possible transmission loss.

Although the LED cannot transmit at speeds as high as the semiconductor laser, it has been widely used for short and intermediate distance communications of small and medium capacity, because the technology is available and costs are lower.

Fig. 2-65. Progress Made in Reducing Transmission Loss

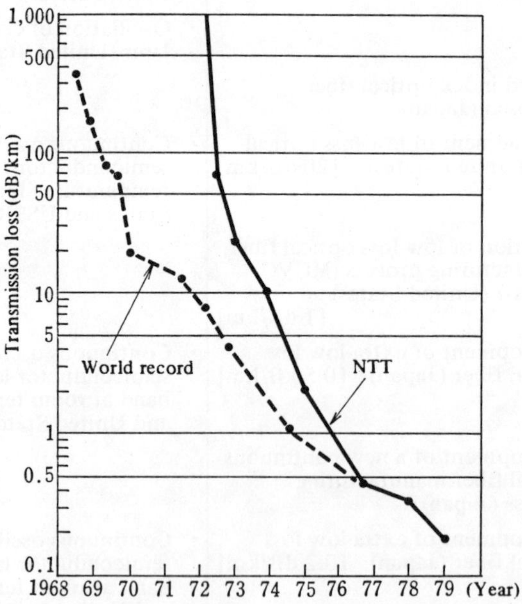

Source: The Nippon Telegraph and Telephone Public Corp. (NTT)

As shown in Table 2-64, research on the semiconductor laser started in Japan in 1957. In 1972, the same year the first low-loss optical fiber was developed, Japan, the United States and the Soviet Union succeeded in continuous laser oscillation at room temperature by using a compound semiconductor made of gallium-aluminum-arsenic (GaAlAs). The optical fiber used then was low in loss in the 0.85μm band (1μm, or micrometer = one millionth of a meter), and because the oscillation wave length of the semiconductor laser coincided with this band, research became very active.

Since then it was discovered that the theoretically low-loss zone for optical fibers lies in a band area of from 1.2 to 1.6μm. Consequently, research on semiconductor materials that oscillate a laser beam within these wave lengths has advanced in parallel with research on improving the reliability of the semiconductor laser itself. As a result, semiconductor lasers have been developed which can oscillate at 0.85 and 1.3μm for more than 100,000 hours continuously at room temperature. Semiconductor lasers that oscillate at 1.55μm, where the transmission loss of quartz fiber becomes the lowest, were achieved by Japan and the United States in 1979.

The light-receiving device must convert even weak light signals into electric signals. The PIN photo diode and the avalanche photo diode (APD) are generally used. The APD is used for high performance systems because of its high sensitivity, high speed, and low noise.

Among APD's, the silicone APD (Si - APD) and the germanium APD (Ge - APD) have been put into practical use. The silicon APD, low in noise and high in performance, is used for short wave lengths (0.85μm) but cannot be used for long wave lengths (1.2 to 1.6μm), where the germanium APD is used although its noise is comparatively large.

What is needed is a light-receiving device with low noise at long wave lengths. Current research is directed toward integration of the PIN photo diode and the FET (field effect transistor), and development of ADP's with new semiconductor materials. Comparison of various kinds of light-receiving devices in the long wave band (Figure 2-66) shows that at the same transmission speed a device with less light-receiving power can more effectively convert weaker light signals into electric signals.

In Europe and the United States, studies are underway on integration of indium-gallium arsenide (InGaAs) and other PIN photo diodes with gallium arsenide FET. In Japan, on the other hand, research is aimed at attaining high-speed transmission of several hundred Mbit/sec., and since integration of the PIN photo diode and FET is not suitable for this purpose, emphasis is being placed on developing an APD with new semiconductor materials.

At present, as indicated in Table 2-67, Japan is leading the world in many aspects of optical communications technology. The country's next targets are development of an optical communications system using a 1.55μm band and integration of optical circuits in light-emitting devices, light-receiving devices, and relay equipment.

Fig. 2-66. Light-Receiving Power vs. Transmission Speed at Long Wave Lengths

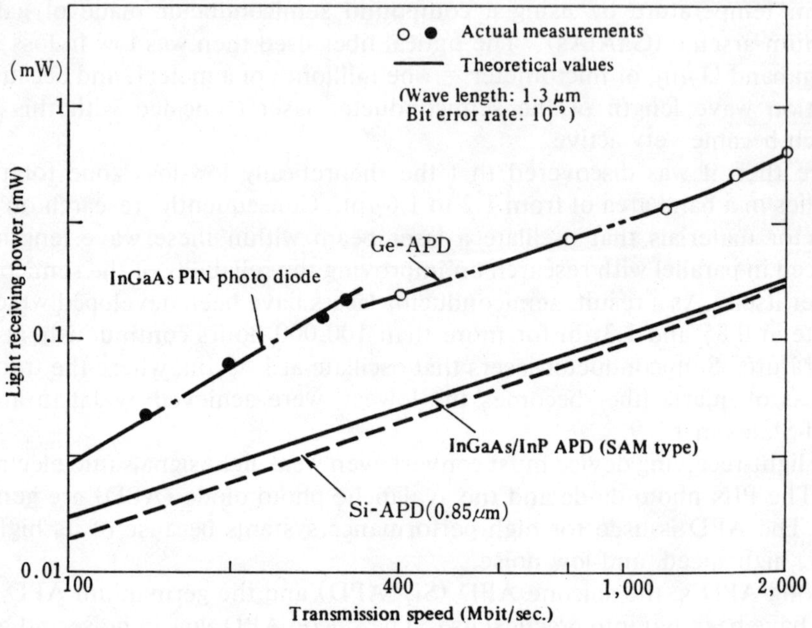

Note: 1 Mbit (megabit)/sec. indicates that 1 million bits of information are transmitted per second

Source: The Opto-electronic Industry and Technology Development Association

Table 2-67. Optical Communications Systems in Various Countries in 1980 (Typical examples of field tests and commercial tests)

Country	Japan		United States	Canada	France	Italy
Total length	18km	18km	11km	50km	7.5km	16km
Number of fiber cores	24, 48	6	36	12	70	18
Transmission speed	6.3Mbit/s 32Mbit/s 100Mbit/s Analog	400Mbit/s	45Mbit/s	274Mbit/s	34Mbit/s	140Mbit/s
Wave length	0.85μm band 1.3μm band	1.3μm band	0.85μm band	0.85μm band	0.85μm band	0.85μm band
Light emitting device	LD (0.85μm) LD, LED (1.3μm)	LD	LD	LD	LD	LD
Light receiving device	Si–APD (0.85μm) Ge–APD (1.3μm)	Ge–APD	Si–APD	Si–APD	Si–APD	Si–APD
Maximum relaying interval	32Mbit/s 10km (0.85μm) 21km (1.3μm) 100Mbit/s 9km (0.85μm) 18km (1.3μm)	30km	6.4km	3.2km	7.5km	8km

Note: LD: Semiconductor laser; LED: Light-emitting diode; APD: Avalanche photo diode

Source: The Science and Technology Agency

(2) Device Technology

Development in semiconductor technology has had immense influence on today's society. The rapid progress toward higher integration of circuits, as shown in Figure 2-68, has made electronic equipment such as computers, calculators, and watches not only light and compact but also of higher performance and reliability.

Fig. 2-68 Integration of IC's

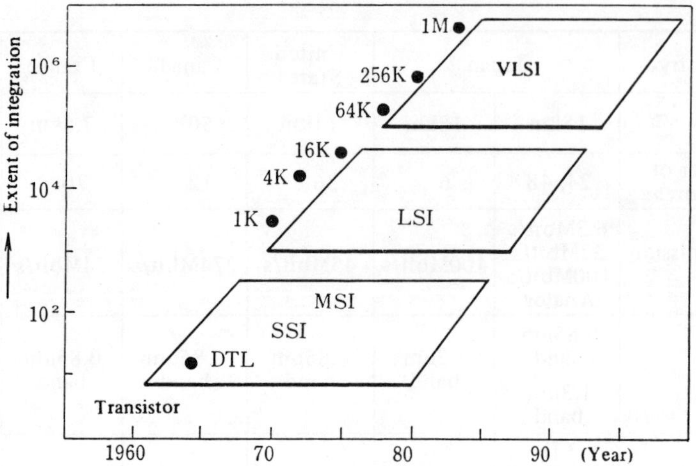

Source: The Electronics Industry Association of Japan

Research in semiconductor technology, because it supports innovations in a variety of fields, is carried out on a national level in many countries, as exemplified by the VLSI (very large scale integration) project in Japan and the VHSI (very high speed integration) project in the United States.

Research goals can be largely classified into higher integration, faster operation speed, and improved stability, each of which will be discussed below.

(a) Higher Integration

Higher integration of IC's has made it possible to put many functions in one device, increase the speed, improve the reliability, and at the the same time drastically cut down the cost per function. Pattern generation is the most important technology in integration. As shown in Figure 2-69, the extent of IC integration has improved hand in hand with progress in pattern generation, as represented here by the decreasing width of the line.

The ultra-violet lithography system was used in the past for pattern generation, but to develop VLSI's such as the 256 kbit RAM (random access memory), a new pattern generation technology became neccessary. Under Japan's VLSI project, an electron beam lithography system was developed with a line width of less than 1 μm. A system of equivalent performance was also developed by NTT, which then used it to successfully develop the world's first 256 Kbit RAM.

Fig. 2-69. Progress in Integration and Pattern Generation

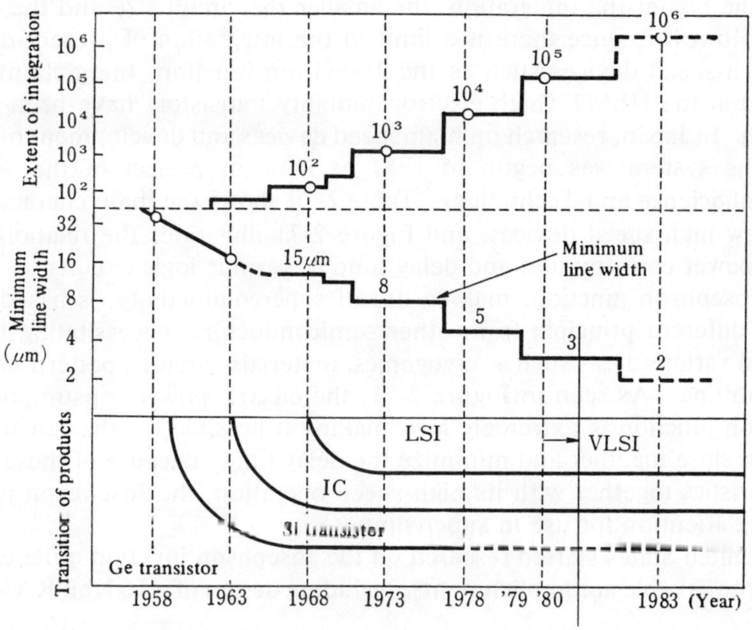

Source: The Japan Society of Electronics Communication

Research on finer pattern generation with methods such as focusing ion beams is also being carried out, but it is generally expected that this technology will reach a limit in the 1990's because of physical restrictions.

Despite such restrictions, various projects are underway to develop methods for still higher integration, especially by multiplying the layers of the IC. The multilayered IC offers high speed processing and very large-scale integration of sensor, signal conversion, memory, and many other functions into one IC. Nevertheless, the technology for controlling the growth of a crystal was a prohibitive factor. However, with the development of a new technique using lasers, research on the multilayered IC has become active, especially in the United States, which successfully test produced a dual-layered CMOS-IC (complementary metal oxide semiconductor integrated circuit). On the other hand, in Japan full-scale research on multilayered IC has just started.

(b) Faster Operation Speed

Along with progress in science and technology, computers are used to process increasingly large amounts of information, and processing speeds about 1,000 times faster than present-day computers have become necessary. To increase the speed, both a parallel processing system and high-speed devices are needed.

Improvements in speed are closely related to increased integration. In other words, the higher the integration, the smaller the circuit size and the faster the speed. However, since there is a limit to the integration of present-day silicon IC's, high-speed devices such as the Josephson junction, the gallium arsenide device, and the HEMT (high electron mobility transistor) have been attracting attention. In Japan, research on high-speed devices and development of a parallel processing system was begun in 1981 as a major project of the Agency of Industrial Science and Technology. Table 2-70 shows the main characteristics of these new high-speed devices, and Figure 2-71 illustrates the relation between electric power consumption and delay time of various logic circuits.

The Josephson junction, making use of superconductivity, is based upon an entirely different principle from other semiconductors, necessitating new technology in various areas such as cryogenics, materials, circuits, pattern generation, and mounting. As seen in Figure 2-71, the electric power consumption of the Josephson junction is extremely low, making it possible for the junctions to be mounted close together and minimize the delay time. Because of these excellent characteristics together with its high-speed operation, the Josephson junction is attracting attention for use in supercomputers.

The United States started research on the Josephson junction quite early, and, with its remarkable accomplishments, including design of a 16 Kbit RAM and test

Table 2-70. Characteristics of High-Speed Devices

New devices / Characteristics	Josephson junction devices (JJ devices)	High electron mobility transistor devices (HEMT devices)	Gallium arsenide devices (GaAs devices)
Operating temperature	Liquid hellium temperature (-269°C)	Liquid nitrogen temperature (-196°C)	Atmospheric temperature
Speed	20~50 times that of silicon devices	10~20 times that of silicon devices	About 5 times that of silicon devices
Electric power consumption	1/1,000 of silicon device or less	1/10~1/20 of silicon device	1/2~1/5 of silicon device
Principle	Tunnel devices using superconductive phenomenon	Using high electron mobility on the boundary layer of laminated semiconductor of GaAs and AlGaAs	Using high electron mobility of GaAs semiconductor

Source: The Agency of Industrial Science and Technology

Fig. 2-71. Electric Power Consumption and Delay Time of Various Logic Circuits

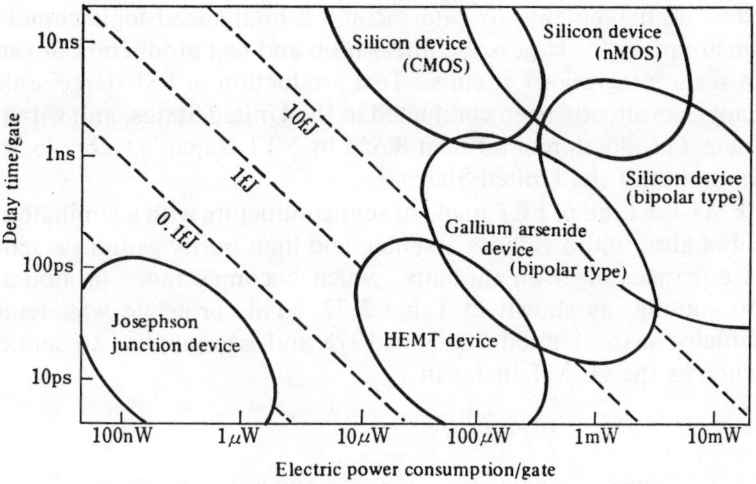

Source: The Agency of Industrial Science and Technology

production of 2 Kbit RAM, is now the world leader in this field.

In Japan, research on the Josephson junction started comparatively late and applications have been made mainly in precision instrumentation. Design of a flat, easily produced Josephson junction and development of a logic circuit with the world's highest operating speed were achieved in Japan, but overall Japan is still behind the United States.

As is clear from Figure 2-71, the galllium arsenide device is inferior to the Josephson junction in performance, but since it can operate at room temperature and is comparatively easy to connect to peripheral circuits, it is also a powerful candidate for use in supercomputers.

The FET and other single-element devices have been used for low noise amplification in earth stations. In this field, gallium arsenide technology in Japan is at the highest level in the world. The integration technology of the gallium arsenide device has advanced rapidly since 1978, based on technology accumulated through development of single-element devices.

It is more difficult to increase the purity of gallium arsenide, a compound material, than silicon. Therefore, progress in the gallium arsenide device depends on development of technology for controlling the concentration of impurities and the thickness of the conducting layer as well as new processing technology suitable for gallium arsenide, which is not resistant against high temperature.

Research in Japan on high integration of gallium arsenide devices has been promoted by the Electrotechnical Laboratory of the Ministry of International Trade and Industry, Nippon Telegraph and Telephone Public Corp. (NTT), and private enterprises. Achievements to date include a high-speed logic circuit with low power consumption for large-scale integration and test production of various MSI (medium-scale integration) circuits. Test production of LSI (large-scale integration) circuits has already been conducted in the United States, and with the recent successful test production of a 1 Kbit RAM by NTT, Japan's technological level is approaching that of the United States.

The HEMT is a kind of FET made of semiconductors with a laminated structure composed of aluminium gallium arsenide and high purity gallium arsenide. This structure enhances electron mobility, which becomes more pronounced when subject to cooling, as shown in Table 2-72. This principle was demonstrated experimentally in the United States in 1978 and later applied to semiconductor devices such as the HEMT in Japan.

Table 2-72. Comparison of Electron Mobility

Devices	Electron mobility (cm^2/Vs)	
	300K	77K
HEMT	6,200	49,300
GaAs FET	4,800	6,500
Silicon MOS FET	570	2,300

Source: The Japan Society of Applied Physics

Recently in Japan, an experiment using a small-scale HEMT device reached an operation speed close to that of the Josephson junction when cooled to liquid nitrogen temperature (about 196°C). Even though the HEMT is operated under extremely low temperature, its temperature is still far higher than that of the Josephson junction. HEMT, because it can operate at room temperature, is expected to be applied in supercomputer development and a wide range of other fields.

The HEMT element, whose operating principle has been confirmed only recently, still has many problems such as crystal growth technology and circuit technology yet to be solved before an LSI circuit can be developed.

In addition, the perfect crystal device, the super-lattice device, and other new devices have also come into the limelight as very high-speed, high-performance devices of the future. The static induction transistor (SIT) proposed in Japan is a kind of a perfect crystal device, whose speed, power and other excellent characteristics have attracted world attention. Research on the super-lattice device, expected to be able to perform operations at speeds far exceeding those of conventional devices, is being actively promoted in the United States. In Japan, both the perfect crystal device and the super-lattice device are being studied as government-promoted projects.

Japan is in a leading position in terms of making quality improvements in existing technology, such as very large-scale integration of silicon semiconductors, but is still behind the United States in the development of new devices.

(c) Improved Stability

Semiconductor devices in common use are built to be used in a comparatively mild environment. In recent years, however, it has become increasingly necessary to use electronic equipment under more severe conditions such as are found in space, nuclear reactors, automobiles, and aircraft. For such purposes, equipment which can withstand radiation, heat, and mechanical impact are needed.

Development of such devices was pioneered by the United States for use in space and for military purposes. One of the targets of the VHSI project in the U.S. is to improve resistance against very intensive radiation and high temperatures.

In Japan, on the other hand, research on such devices has just started. The initial target has been to heighten the impact resistance of silicon semiconductors and develop a device with integration equivalent to the microprocessor and sufficient resistance against radiation and heat to be mounted on artificial satellites. In addition, radiation-resistant gallium arsenide devices for nuclear reactors and heat-resistant silicon carbide (SiC) devices for automobile engines are being developed.

CHAPTER 3

RESEARCH EXPENDITURES AND PERSONNEL

1. OVERVIEW

(1) Expenditures

A country's total research expenditures, both public and private, constitute one of the main indices of trends in research activities. In Japan, research expenditures for the 1980 fiscal year totalled $18.74 billion, 15.3 percent over the figure for the previous year, $16.25 billion. Real research expenditures, adjusted for inflation, leveled off in 1974 and 1975, but then rose from 1976 on, as shown in Figure 3-1.

Internationally, research expenditures in the U. S. and the Soviet Union are head and shoulders above other leading nations, followed by West Germany and Japan, as shown in Figure 3-2. Comparing rates of growth in national expenditures, however, Figure 3-3 shows that Japan is second only to West Germany in its growth rate for real expenditures for research.

Next, the ratio of research expenditures to national income in several leading nations will be examined. In Japan, as shown in Figure 3-4, this ratio rose steadily year by year, reaching 2.16 percent in 1974. It then leveled off until 1979, when it started rising again and attained a record high 2.42 percent in 1980.

Fig. 3-1. Changes in Research Expenditures in Japan

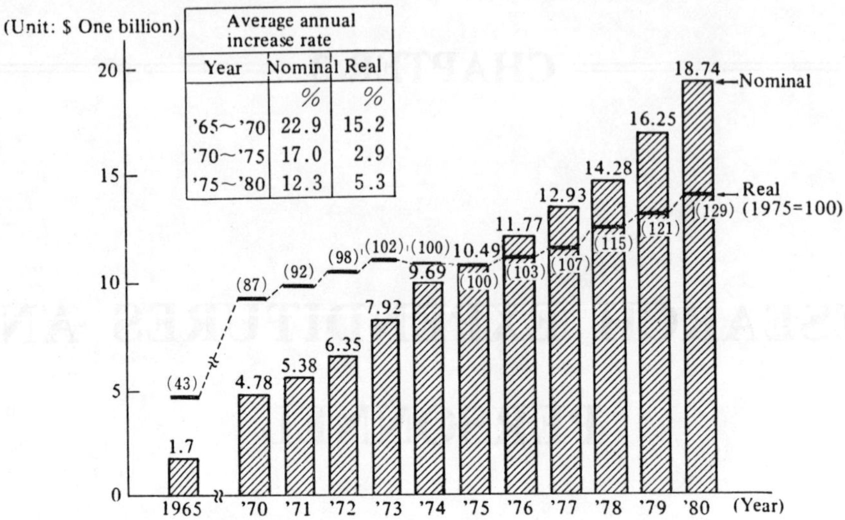

Source: The Statistics Bureau, Prime Minister's Office

Fig. 3-2. Changes in Research Expenditures of Leading Nations

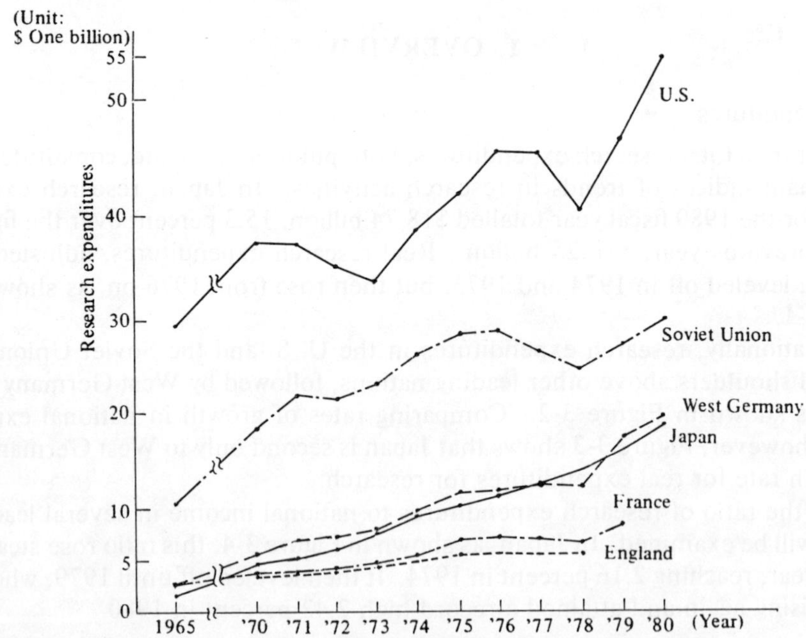

Source: The Statistics Bureau, Prime Minister's Office

Fig. 3-3. Rates of Growth in Real Research Expenditures in Leading Nations

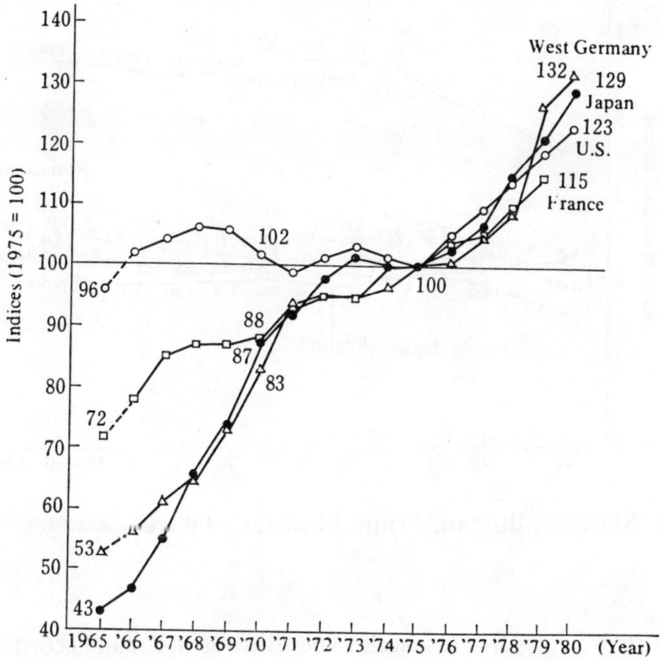

Source: The Statistics Bureau, Prime Minister's Office

In 1977, the Council for Science and Technology, taking into consideration the need for future investments and trends in Europe and the United States, announced a figure of 2.5 percent as an immediate target for the ratio of research expenditures to national income, and also set 3.0 percent as a long-term target. Active investments in recent years have moved the country a step closer to attainment of this target.

The same figure shows that in all of the leading nations this ratio has generally been on the rise. At present, the ratio for Japan is higher than that of France, but is still at a low level compared to other countries.

Fig. 3-4. Ratio of Research Expenditures to National Incomes

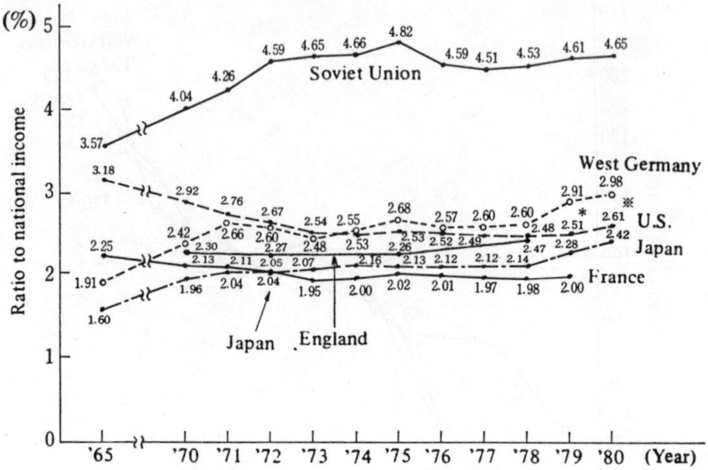

Source: The Statistics Bureau, Prime Minister's Office

Next, Figure 3-5 classifies research expenditures in Japan according to the kind of organization making them. In 1980, corporations accounted for the biggest share, standing at $12.57 billion, or 67.1 percent of the total, an increase in value of 17.9 percent over the previous year. Universities came next at $3.3 billion, or 17.6 percent of the total, up 5.9 percent followed by research laboratories at $2.87 billion, or 15.3 percent of the total, up 15.6 percent.

As seen in Figure 3-6, during the latter half of the 1960's, a period of high economic growth in Japan, total research expenditures grew sharply because of increases made by corporations, which account for a large proportion of the total. However, in the first half of the 1970's, the annual growth rate for corporate research expenditures fell off, actually recording negative growth rates in 1974 and 1975. At the same time, the annual growth rate of total research expenditures also decreased, dropping to minus 1.9 percent in 1974. However, from 1976 onwards, corporate research expenditures started on an upward trend, and the annual growth rate of total research expenditures began to increase as well, marking 6.8 percent in 1980.

Fig. 3-5. Changes in Research Expenditures by Organization

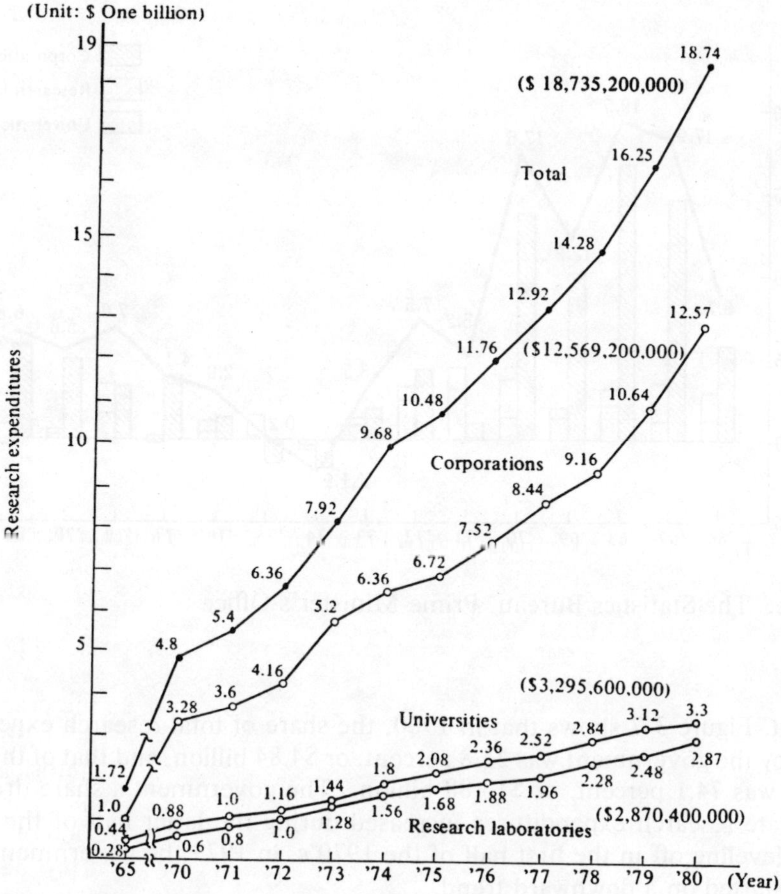

Source: The Statistics Bureau, Prime Minister's Office

Fig. 3-6. Annual Growth Rate of Research Expenditures and Share Contributed by Organization

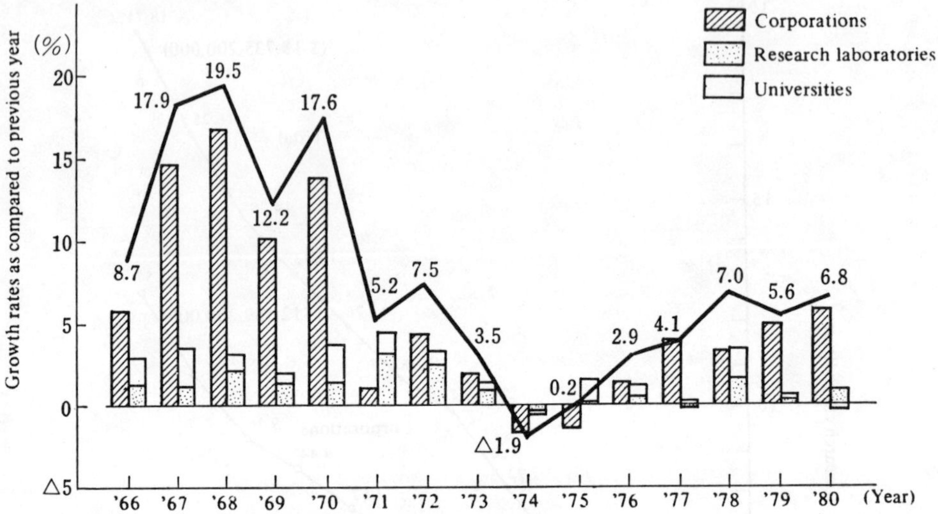

Source: The Statistics Bureau, Prime Minister's Office

Next, Figure 3-7 shows that in 1980, the share of total research expenditures made by the government was 25.8 percent, or $4.84 billion, and that of the private sector was 74.1 percent, or $13.88 billion. The government's share dropped as corporate research expenditures increased during the latter half of the 1960's. After leveling off in the first half of the 1970's, in 1979 the government's share again started on a downward trend.

As shown in Figure 3-8, the large majority of research expenditures funded by the government are used by public research laboratories, semi-government laboratories, or universities, and only a very small portion goes to the private sector.

Fig. 3-7. Changes in Public and Private Shares of Research Expenditures

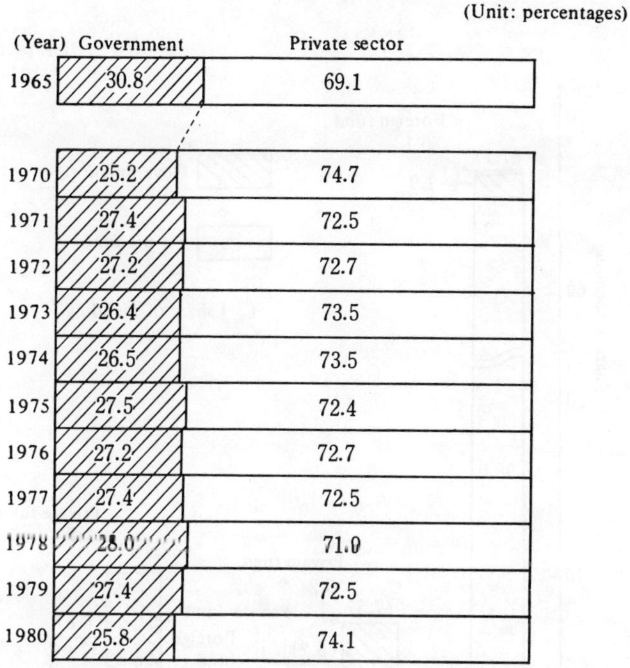

Notes: 1. Government refers to national and local public bodies.
2. Each year, there are research expenditures from abroad of approximately 0.1 percent.

Source: The Statistics Bureau, Prime Minister's Office

Fig. 3-8. Share of Funding of Research Expenditures at Corporation, Research Laboratories, and Universities

(Unit: percentages)

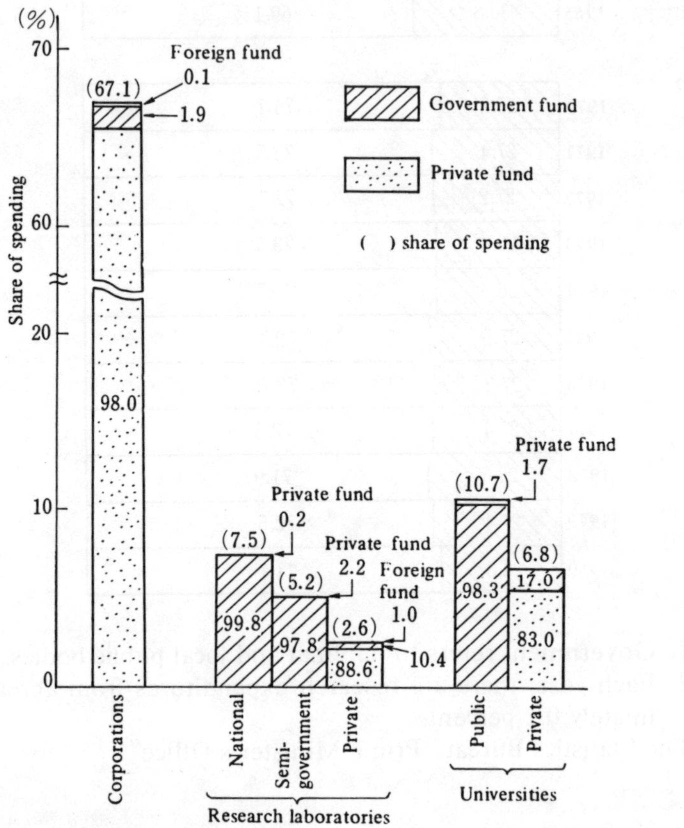

Source: The Statistics Bureau, Prime Minister's Office

As seen in Table 3-9, expenditures by the industrial sector in Japan account for approximately two-thirds of all domestic research expenditures, approximately the same as in other nations. However, of the proportion funded by the government, the defense-related budget is comparatively low in Japan. Differences in tax systems make simple comparisons difficult, but defense-related research spending was 47.9 percent in the U. S. (1980), 48.1 percent in England (1978), 46.8 percent in West Germany (1979), and 51.1 percent in France (1979), all of which are larger than Japan's 25.8 percent (1980).

Table 3-9. Breakdown by Organization of Research Expenditures in Leading Nations

(Unit: percentages)

Country \ Sector	Industrial	Government	Non-profit research laboratories	Universities
Japan (1978)	64.2	13.6	2.3	20.0
Japan (1979)	65.6	13.4	1.9	19.1
Japan (1980)	67.1	12.7	2.6	17.6
U.S. (1980)	69.9	13.2	3.6	13.3
England (1978)	64.2	21.2	3.2	11.4
West Germany (1979)	65.1	17.0	—	17.9
France (1979)	59.5	23.6	1.4	15.5

Notes: 1. Humanities and social science research expenditures are included in the figures for all countries except for Japan.

2. Share of non-profit research institutions in West Germany are inoluded under "Government"

Source: The Statistics Bureau, the OECD, and the U. S. National Science Foundation

Table 3-10. Share of Research Expenditures Funded by Government

Country	Classification	Government fund	Government fund excluding defense-related research expenditures
Japan	(1978)	28.0 %	27.5 %
Japan	(1979)	27.4	26.9
Japan	(1980)	25.8	25.4
U.S.	(1980)	47.9	33.2
England	(1978)	48.1	31.6
West Germany	(1979)	46.8	43.6
France	(1979)	51.1	37.9

Note: Proportion funded by government after excluding defense-related research expenditures is calculated as follows:

$$\frac{\text{Government research expenditures} - \text{Defense-related research expenditures}}{\text{Research expenditures} - \text{Defense-related research expenditures}} \times 100$$

Source: The Statistics Bureau, the OECD, and the U. S. National Science Foundation

Excluding defense-related expenditures, the proportion of research funded by government is 43.6 percent for West Germany, 37.9 percent for France, 33.2 percent for the U. S., 31.6 percent for England, and 25.4 percent for Japan, as shown in Table 3-10.

Research expenditures include expenses for personnel, raw materials, tangible fixed assets (land, buildings, equipment, etc.) and other items. Figure 3-11 shows that of these items, in Japan the proportion taken by personnel expenses is the highest. From 1970 onwards, this proportion tended to rise until 1975, when it started to drop, reaching 46.9 percent in 1980. On the other hand, expenditures for the purchase of tangible fixed assets, previously in a downward trend, in that year started to increase, reaching 18.3 percent, while raw materials expenditures also rose somewhat to 15.5 percent. Following personnel expenditures, the next

Fig. 3-11. Breakdown of Research Expenditures by Item

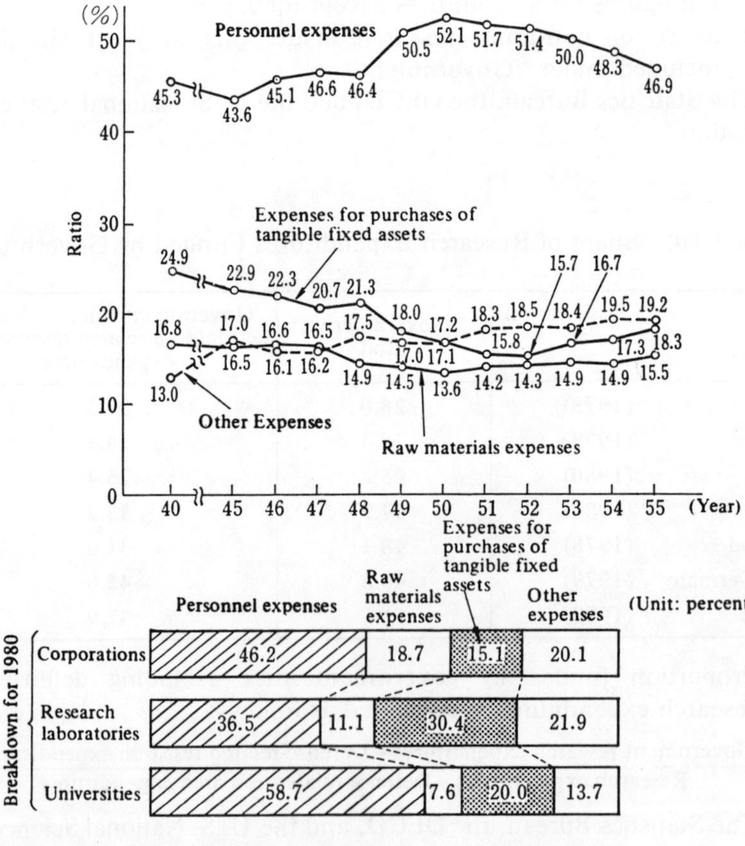

Source: The Statistics Bureau, Prime Minister's Office

largest item, taking 19.2 percent of the total, is here referred to as other expenditures required for research, and includes expenditures for publications, clerical assistance, travel and communications.

Next, Figure 3-12 shows that in Japan the share of expenditures for basic research decreased somewhat in 1980, while that for developmental research increased. Furthermore, the share of developmental research at corporations is extremely high, while in universities much weight is placed upon basic and applied research.

The share of developmental research is relatively large in the U. S., while that of basic research is large in West Germany and France, as shown in Figure 3-13. Japan and England have proportions midway between these countries.

Fig. 3-12. Expenditures in Japan by Type of Research

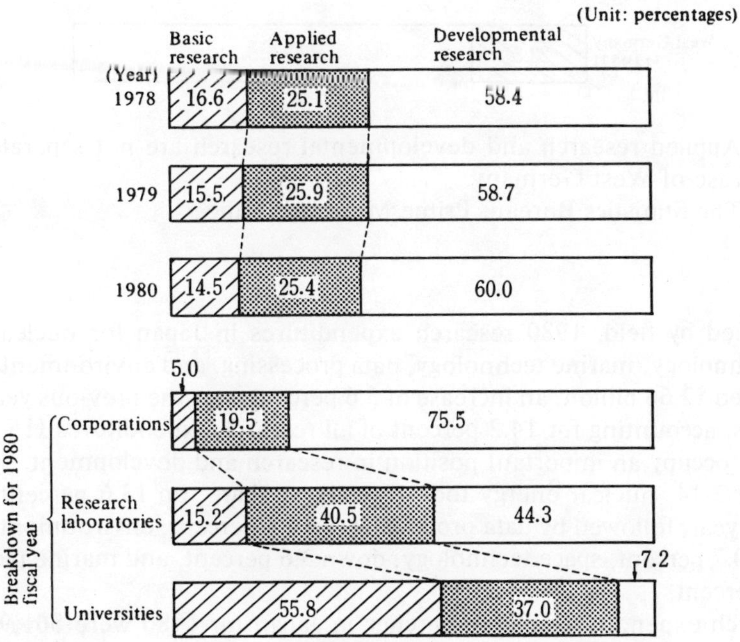

Source: The Statistics Bureau, Prime Minister's Office

Fig. 3-13. Expenditures by Type of Research

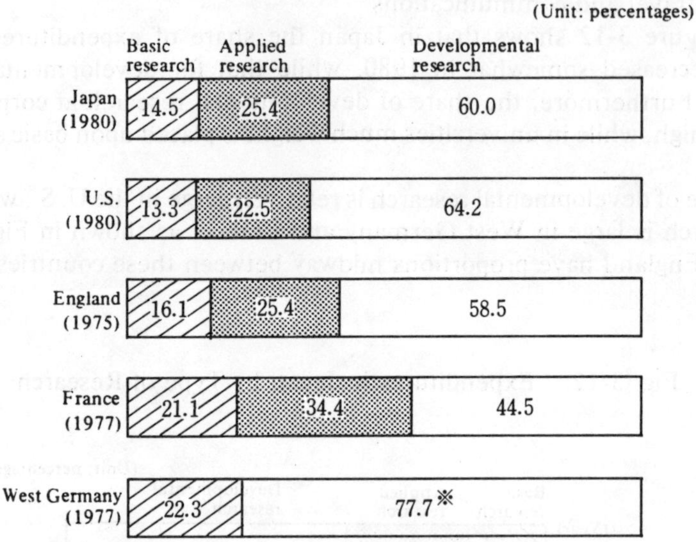

Note: Applied research and developmental research are not separated in the case of West Germany.

Source: The Statistics Bureau, Prime Minister's Office

Classified by field, 1980 research expenditures in Japan for nuclear energy, space technology, marine technology, data processing, and environmental protection totaled $2.66 billion, an increase of 5.6 percent over the previous year. These five fields, accounting for 14.2 percent of all research expenditures (15.5 percent in 1979), occupy an important position in research and development. As shown in Figure 3-14, nuclear energy took the largest share, up 13.6 percent over the previous year, followed by data processing, up 4.4 percent, environmental protection, up 0.7 percent, space technology, down 3.6 percent, and marine technology, up 8.6 percent.

Research expenditures per researcher in Japan for 1980 were $61,920, a 7.4 percent increase over the previous year. As shown in Figure 3-15, although nominal research expenditures per researcher have risen every year recently, in real terms, due to inflation, the figures for the latter half of the 1970's leveled off, and in 1980 dropped 0.5 percent to $44,800, comparable to the level in the latter half of the 1960's. Expenditures per person at research laboratories were the highest, at $100,240, followed by corporations, at $72,560, and universities, at $32,720. Rates of increase over the previous year were similar, with research laboratories the highest at 12.6 percent, followed by corporations at 7.1 percent

Fig. 3-14. Research Expenditures by Field of Research

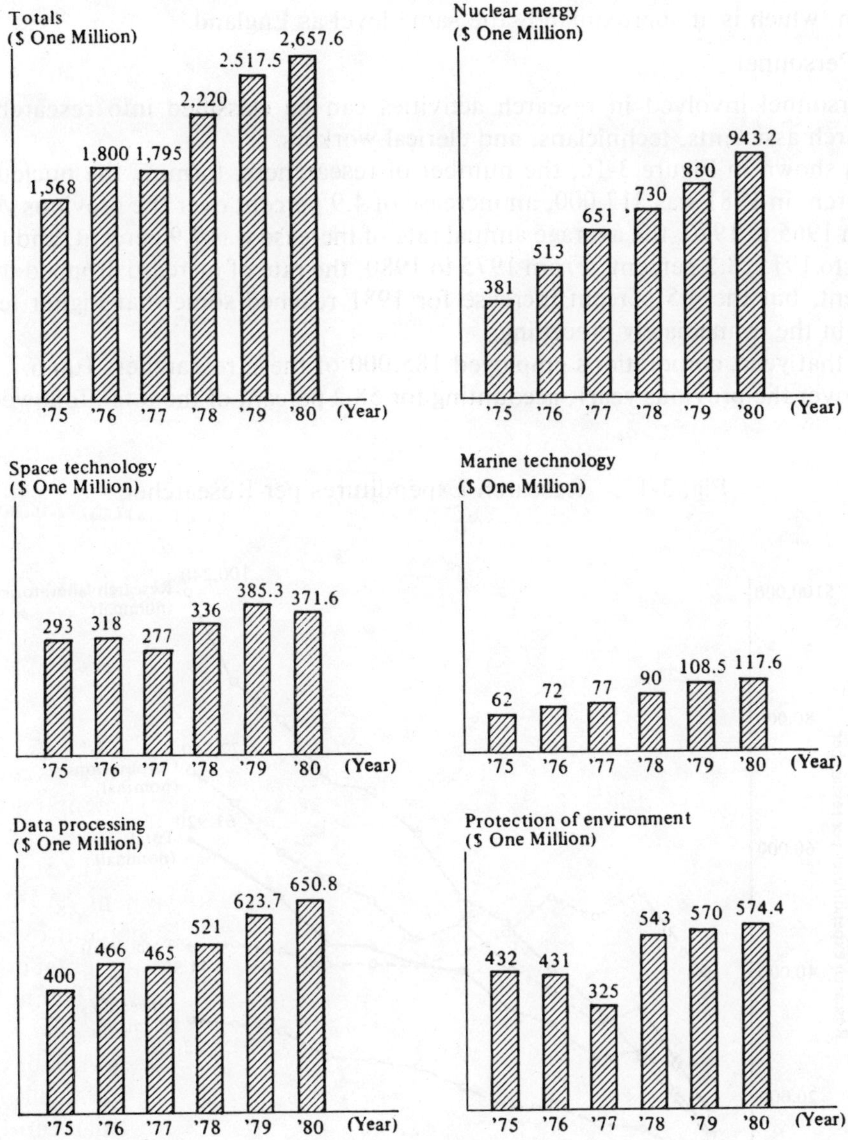

Source: The Statistics Bureau, Prime Minister's Office

and universities at 1.7 percent. Although not shown in the figure, research expenditures per researcher in West Germany and France are approximately double the figure for Japan, while the figure for the U. S. is about 1.4 times that of Japan, which is at approximately the same level as England.

(2) Personnel

Personnel involved in research activities can be classified into researchers, research assistants, technicians, and clerical workers.

As shown in Figure 3-16, the number of researchers, forming the nucleus of research, in 1981 was 317,000, an increase of 4.9 percent over the previous year. From 1965 to 1970, the average annual rate of increase was 7.9 percent, and from 1970 to 1975, 8.2 percent. From 1975 to 1980, the rate of increase dropped to 3.5 percent, but the 4.5 percent increase for 1981 reached somewhat higher levels than in the immediately preceding.

In that year, corporations employed 185,000 of these researchers (up 6.7 percent over the previous year), accounting for 58.2 percent of the total, followed by

Fig. 3-15. Research Expenditures per Researcher

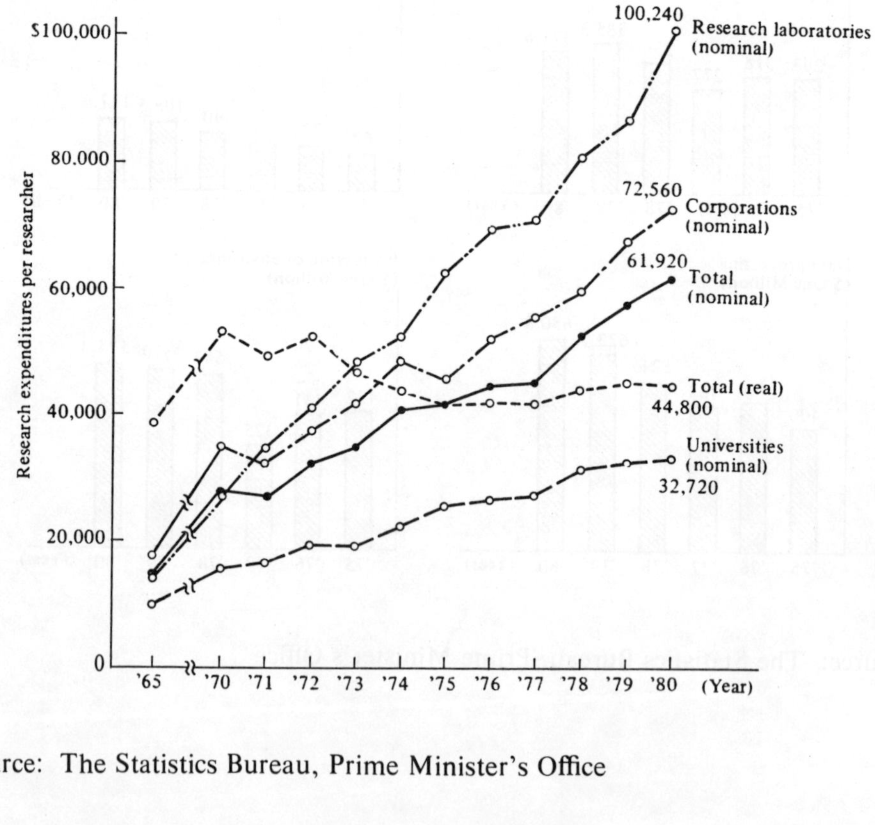

Source: The Statistics Bureau, Prime Minister's Office

universities at 103,000 (up 1.9 percent) or 32.3 percent, and research laboratories, at 30,000 (up 4.8 percent) or 9.5 percent.

Comparing the number of researchers in Japan with other leading nations, Figure 3-17 shows that the Soviet Union, with 1.37 million researchers, has the largest number, followed by the U. S. with 644,000, respectively 4.3 and 2.0 times as many as in Japan. However, the number of researchers in West Germany, England, and France is far less than in Japan. In terms of numbers of researchers per 10,000 population, Japan is roughly on the same level as the U. S.

Fig. 3-16. Changes in Number of Researchers in Japan

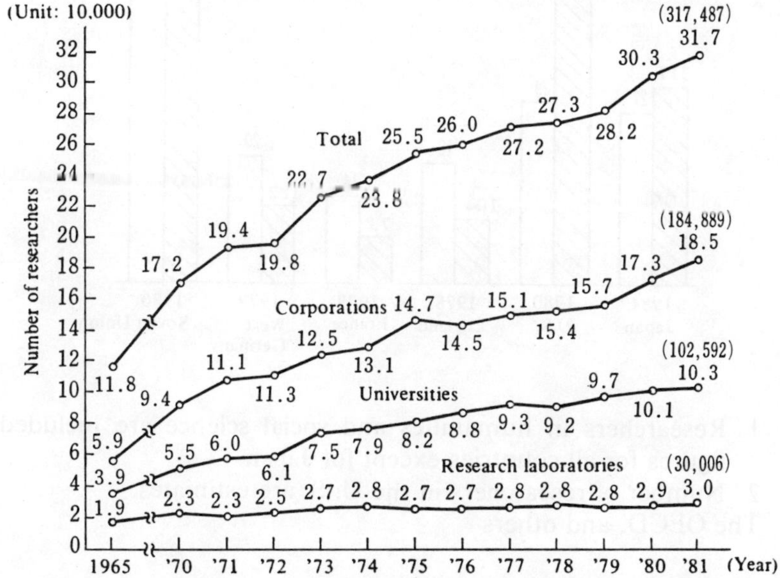

Source: The Statistics Bureau, Prime Minister's Office

Fig. 3-17. Number of Researchers in Leading Nations

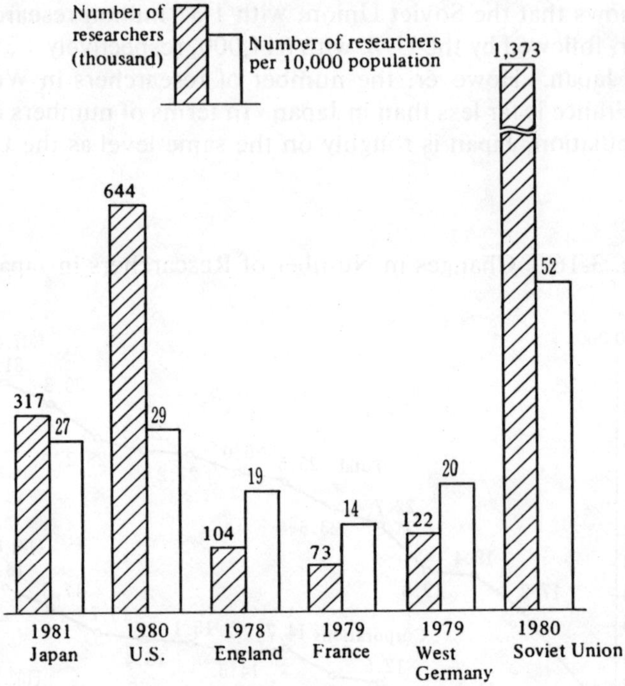

Notes: 1. Researchers in humanities and social science are included in the figures for all countries except for Japan.

2. Number of researchers in the U. S. are estimates.

Source: The OECD, and others

As shown in Figure 3-18, in 1981 in Japan the total number of research-related personnel was 548,000, an increase of 5.2 percent over the previous year. Figure 3-19 shows that the proportion of researchers in the total for that year was 57.9 percent, that of research assistants 14.6 percent, that of technicians 16.3 percent, and that of clerical workers 11.2 percent. Fianally, Figure 3-20 shows that the number of research assistants, technicians, and clerical workers per researcher has been decreasing for a number of years, leveling off at 0.73 persons, fractionally down from the previous year.

Fig. 3-18. Changes in Number of Research-Related Personnel

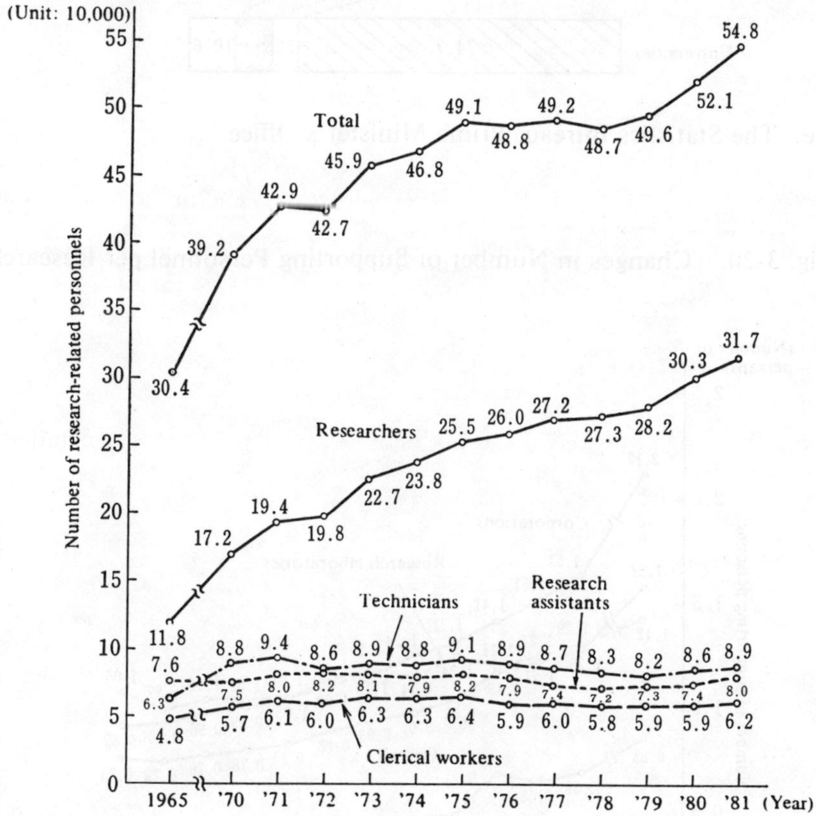

Source: The Statistics Bureau, Prime Minister's Office

Fig. 3-19. Breakdown of Research-Related Personnel (1981)

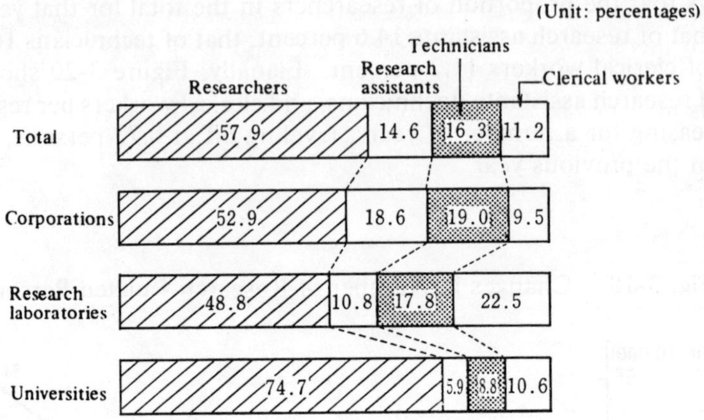

Source: The Statistics Bureau, Prime Minister's Office

Fig. 3-20. Changes in Number of Supporting Personnel per Researcher

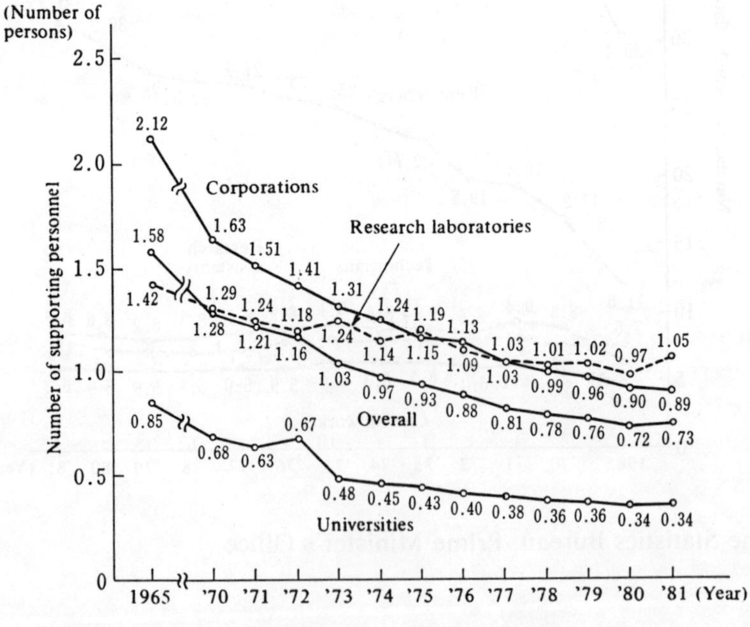

Source: The Statistics Bureau, Prime Minister's Office

2. CORPORATIONS

In Japan, research expenditures at corporations were $12.57 billion in 1980, a 17.9 percent increase over the previous year, and accounted for 67.1 percent of the country's total research expenditures. Of these corporate expenditures 75.5 percent was allocated for developmental work leading to new products and new manufacturing processes or improvements in existing technology.

Table 3-21. Breakdown of Corporate Research Expenditures by Industry

(1980)

Industry	Number of companies conducting research	Ratio (%)	Amounts of research expenditures (Unit: $ One million)	Ratio (%)
All industries	17,468	100.0	12,569.024	100.0
Agriculture and fisheries	143	0.8	15.416	0.1
Mining	240	1.4	57.132	0.5
Construction	1,372	7.9	302.760	2.4
Manufacturing	15,613	89.4	11,582.284	92.1
Food	1,904	10.9	360.948	2.9
Textile	825	4.7	136.836	1.1
Pulp and paper	219	1.3	63.696	0.5
Publishing and printing	113	0.6	22.940	0.2
Chemical	1,808	10.4	2,233.008	17.8
Petroleum and coal	249	1.4	259.624	2.1
Rubber	230	1.3	203.520	1.6
Ceramics	1,328	7.6	322.888	2.6
Iron and steel	273	1.6	588.256	4.7
Non-ferrous metals	230	1.3	219.836	1.7
Metal products	1,902	10.9	208.108	1.7
Machinery	2,248	12.9	875.508	7.0
Electrical machinery	1,762	10.1	3,268.896	26.0
Transportation machinery	374	2.1	2,041.816	16.2
Precision machinery	589	3.4	397.352	3.2
Others	1,559	8.9	369.052	2.9
Transportation, communications, and public utilities	100	0.6	611.432	4.9

Note: Companies capitalized at $20,000 or more
Source: The Statistics Bureau, Prime Minister's Office

The number of companies capitalized at twenty thousand dollars or more and conducting research activities amounted to 17,000 as of 1980, of which the manufacturing industry, at 89.4 percent, accounted for the large majority, followed by the construction industry, as shown in Table 3-21.

The manufacturing industry as a whole used 92.1 percent of total corporate research expenditures. Within the manufacturing industry, the electrical machinery industry spent 26.0 percent, the chemical industry, 17.8 percent, and the transportation machinery industry, 16.2 percent, these three together accounting for 60 percent of total corporate research expenditures. From 1970 to 1980, the average annual growth rate in research expenditures was a very high 18.9 percent for the transportation machinery industry, followed by 14.0 percent for the electrical machinery industry and 12.6 percent for the chemical industry.

The ratio of research expenditures to sales, if high, indicates strong interest in research. The ratio for industry as a whole was 1.05 percent for 1965. In the following years it gradually increased, peaking at 1.57 percent in 1978. It then dropped, reaching 1.48 percent in 1981. For that year, the electrical machinery industry had the highest ratio at 3.71 percent, followed by the precision machinery industry at 3.02 percent, as shown in Figure 3-22.

As for the make-up of corporate research expenditures, in the past fifteen years or so, as shown in Figure 3-23, personnel expenses accounted for approximately 50 percent of corporate research expenses.

Figure 3-24, reviewing changes in the kind of research in which corporations invested, shows that from 1965 to 1975 the share of expenditures for both basic and applied research fell while the share for developmental research consistently rose. From 1975 onwards, however, the ratio has been more or less constant, and in 1980 was 5.0 percent for basic research, 19.5 percent for applied research, and 75.5 percent for developmental research.

Fig. 3-22. Changes in Ratio of Research Expenditures to Sales

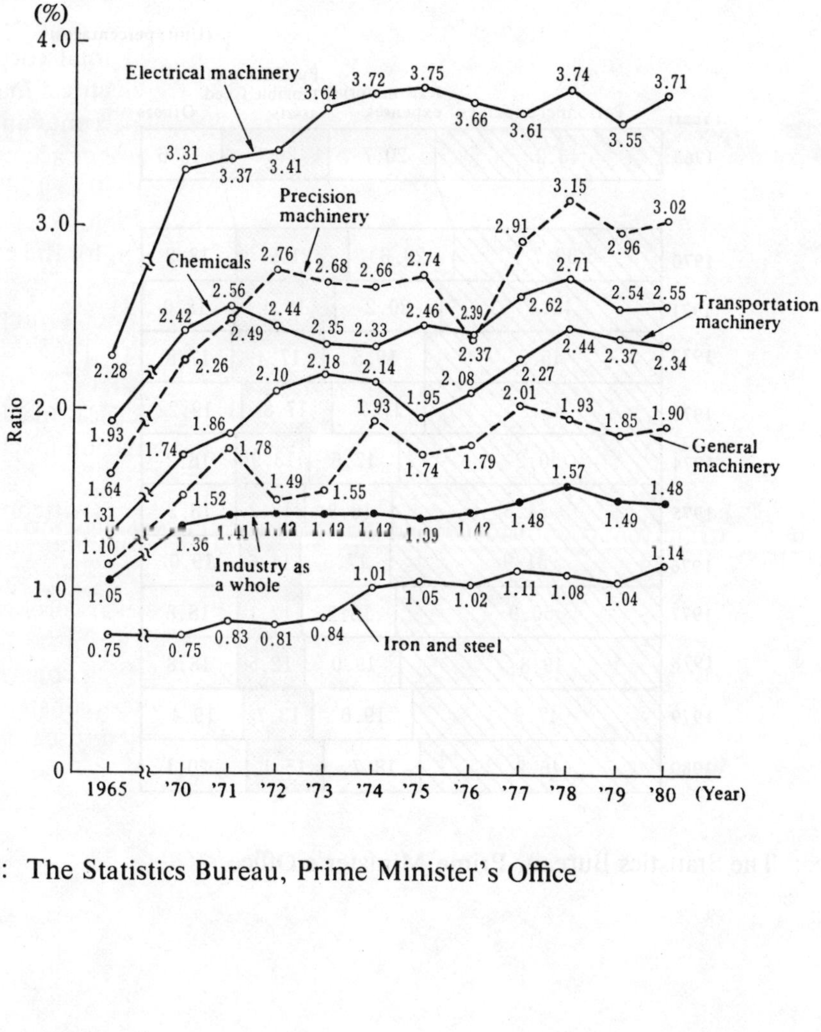

Source: The Statistics Bureau, Prime Minister's Office

Fig. 3-23. Changes in Research Expenditures at Corporations Broken Down by Item

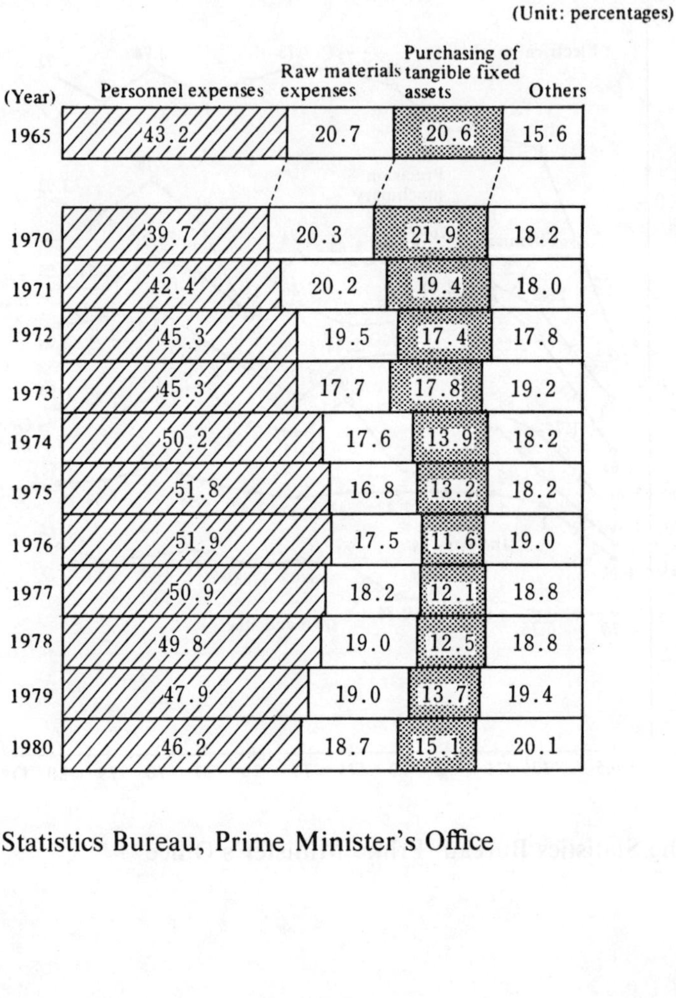

Source: The Statistics Bureau, Prime Minister's Office

Fig. 3-24. Changes in Research Expenditures at Companies by Type of Research

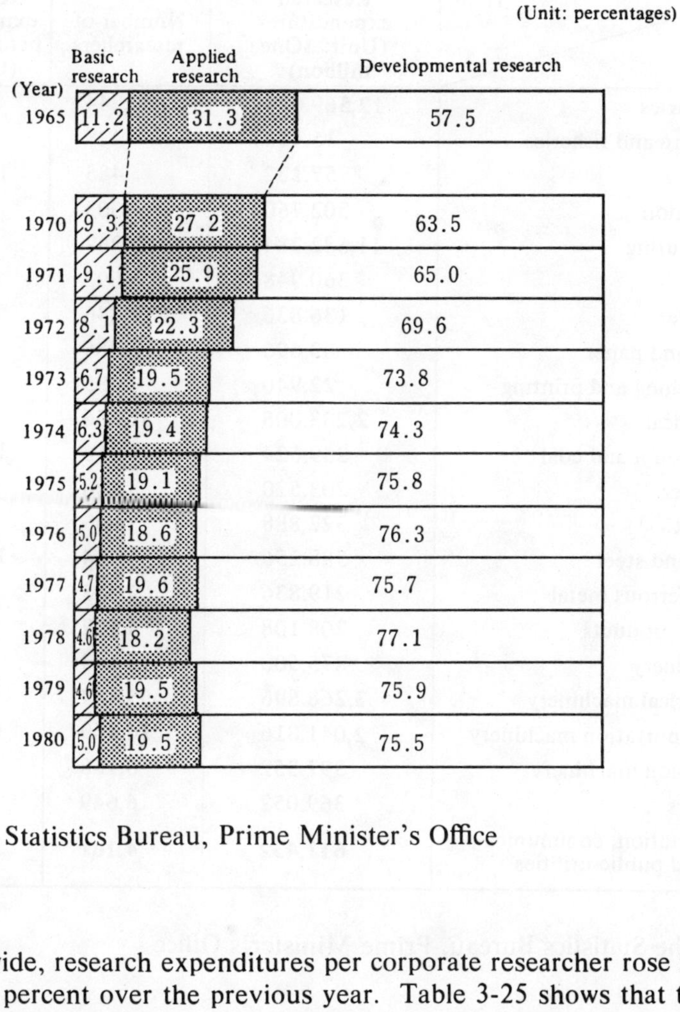

Source: The Statistics Bureau, Prime Minister's Office

Industry-wide, research expenditures per corporate researcher rose in 1980 to $72,560, 7.1 percent over the previous year. Table 3-25 shows that the largest sum, $197,280, was in the petroleum and coal industry, followed by the iron and steel industry and the transportation machinery industry.

Table 3-25. Research Expenditures at Corporations per Researcher (1980)

Industry \ Item	Research expenditures (Unit: $One million)	Number of researchers	Research expenditures per researcher (Unit: $)
All industries	12,569.024	173,244	72,560
Agriculture and fisheries	15.416	428	36,000
Mining	57.132	485	117,800
Construction	302.760	4,303	70,360
Manufacturing	11,582.284	163,867	70,680
Food	360.948	6,092	59,240
Textile	136.836	3,846	35,560
Pulp and paper	63.696	1,355	47,000
Publishing and printing	22.940	485	47,280
Chemical	2,233.008	31,556	70,760
Petroleum and coal	259.624	1,316	197,280
Rubber	203.520	3,180	64,000
Ceramics	322,888	5,355	62,160
Iron and steel	588.256	4,434	132,680
Non-ferrous metal	219.836	2,385	92,160
Metal products	208.108	4,117	50,560
Machinery	875.508	15,273	57,320
Electrical machinery	3,268.896	55,467	58,920
Transportation machinery	2,041.816	16,169	126,280
Precision machinery	397.352	6,188	64,200
Others	369.052	6,649	55,520
Transportation, communications, and public utilities	611.432	4,161	146,960

Source: The Statistics Bureau, Prime Minister's Office

The number of researchers at corporations, shown in Table 3-26, in 1981 was 185,000, up 6.7 percent over the previous year. The manufacturing industries, with 175,000 researchers, accounts for 94.7 percent of the total. Among them, just two, the electrical machinery industry (with 59,000, or 31.8 percent) and the chemical industry (with 33,000, or 17.8 percent) together account for half of the corporate total. In addition, although for all industries the average number of researchers per 10,000 employees was 277 that year, the average for manufacturing industries alone was 348. Other industries with higher than average figures are the electrical machinery industry, the chemical industry, the precision machinery industry, and the petroleum and coal industry.

Table 3-26. Numbers of Researchers at Corporations by Industry (1981)

Industry / Item	Number of researchers	Proportion	Number of employees	Number of researchers per 10,000 employees
All industries	184,889	100.0	6,684,668	277
Agriculture and fisheries	300	0.2	38,535	78
Mining	637	0.3	54,632	117
Construction	4,829	2.6	436,396	111
Manufacturing	175,088	94.7	5,031,229	348
Food	8,342	4.5	436,757	191
Textile	2,371	1.3	189,877	125
Pulp and paper	1,527	0.8	102,244	149
Publishing and printing	477	0.3	71,582	67
Chemical	32,847	17.8	558,599	588
Petroleum and coal	2,363	1.3	66,325	356
Rubber	3,423	1.9	110,074	311
Ceramics	4,799	2.6	230,071	209
Iron and steel	4,800	2.6	338,509	142
Non-ferrous metals	2,895	1.6	119,387	242
Metal products	4,348	2.4	191,749	227
Machinery	15,390	8.3	515,424	299
Electrical machinery	58,873	31.8	966,604	609
Transportation machinery	17,682	9.6	643,356	275
Precision machinery	7,061	3.8	174,247	405
Others	7,890	4.3	316,424	249
Transportation, communications and public utilities	4,035	2.2	1,123,876	36

Source: The Statistics Bureau, Prime Minister's Office

Outlining the ratio of researchers classified by speciality, Figure 3-27 shows that engineering has the highest at 58.8 percent, followed by physical science at 30.4 percent, these two fields together accounting for three-quarters of the total. Within engineering, specialists in such fields as machinery, shipbuilding, aeronautics, electrical engineering, and communications take up the majority, while in physical science the figure for specialists in chemicals is high.

Fig. 3-27. Proportions of Researchers at Corporations by Speciality (1981)

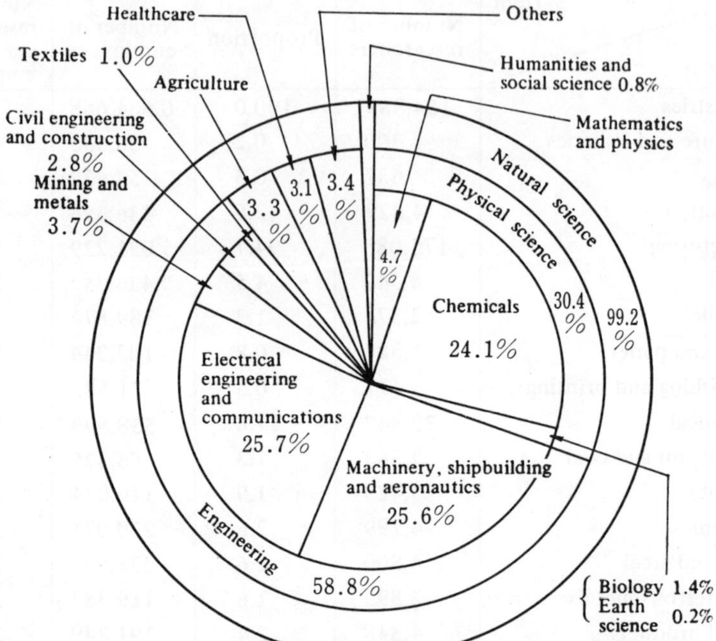

Source: The Statistics Bureau, Prime Minister's Office

3. RESEARCH LABORATORIES

Research laboratories in Japan carry out both large-scale work such as nuclear energy and space development and research on behalf of those industries in which independent research is very difficult. Since it is difficult for the private sector to carry out this work, research laboratories are largely managed and funded by the government.

Research expenditures at research laboratories for the 1980 fiscal year were \$2.87 billion, a 15.6 percent increase over the previous year, and accounted for 15.3 percent of Japan's total research expenditures. The government funded 83.3 percent of these expenditures.

As shown in Table 3-28, from 1979 to 1980 the number of national laboratories increased to 89, the number of public laboratories increased to 638, the number of private laboratories increased to 256, and the number of semi-government laboratories, such as the Japan Atomic Energy Institute, increased to nine. As a

Table 3-28. Changes in Number of Research Laboratories

Classification \ Fiscal year	1965	1970	1975	1976	1977	1978	1979	1980
Totals	845	831	884	915	905	894	927	992
National	83	88	84	83	85	83	87	89
Public	560	551	585	588	598	585	594	638
Private	196	183	209	238	216	218	239	256
Semi-government	6	9	6	6	6	8	7	9
Physical science	95	62	101	102	102	99	114	126
Engineering	230	234	259	276	270	263	282	298
Agriculture	408	409	416	418	418	416	418	450
Healthcare	107	110	108	119	115	116	113	118
Others	5	16	–		–	–	–	–

Source: The Statistics Bureau, Prime Minister's Office

result, in 1980 national laboratories accounted for nine percent, public laboratories, 64.3 percent, private laboratories, 25.8 percent, and semi-government laboratories, 0.9 percent.

National research laboratories are mostly involved in the study of engineering (32.6 percent) and agriculture (32.6 percent), and public laboratories in agriculture (59.4 percent) and engineering (22.6 percent). At private laboratories, engineering (47.7 percent) accounts for the largest share, followed by health care (22.7 percent). Most research at semi-government laboratories is in physical science and engineering.

Research expenditures for 1980 were $742 million for national laboratories, $664 million for public laboratories, $490 million for private laboratories, and $975 million for semi-government laboratories, accounting for 25.8 percent, 23.1 percent, 17.1 percent, and 34.0 percent respectively of total research laboratory expenditures, as shown in Figure 3-29.

Fig. 3-29. Changes in Research Expenditures at Research Laboratories

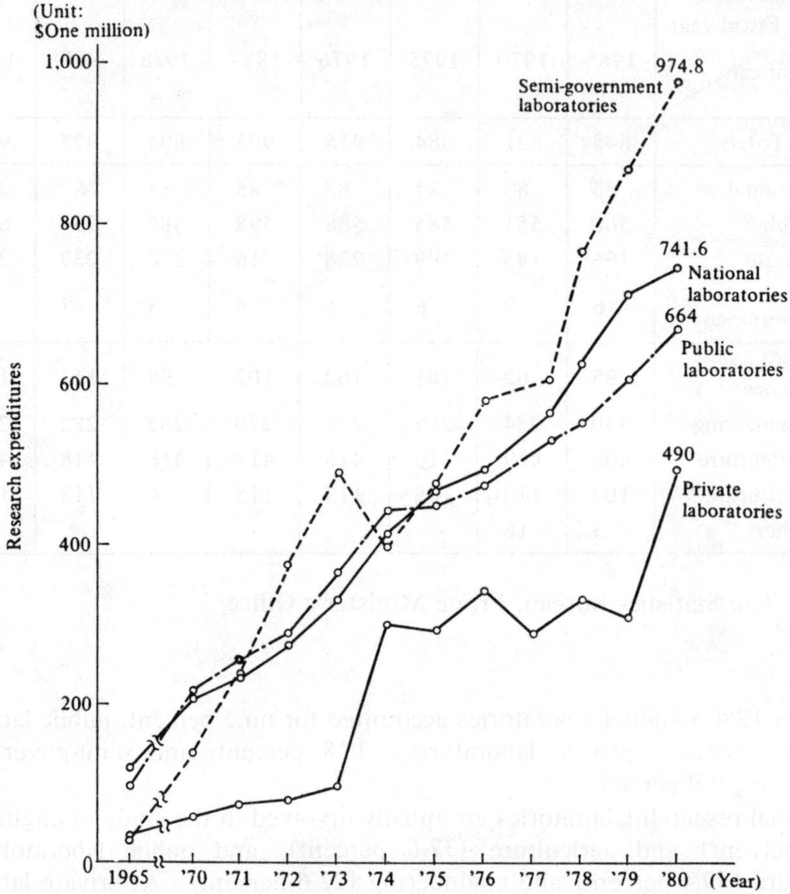

Note: Research expenditures in 1974 dropped at semi-government laboratories but rose at private laboratories because mining technoloy research groups were reclassified from semi-government to private laboratories.

Source: The Statistics Bureau, Prime Minister's Office

Figure 3-30 breaks down expenditures by speciality, and Figure 3-31 by expense. At public laboratories the proportion taken up by personnel expenses, and at semi-government laboratories expenses for the purchasing of tangible fixed assets, are extremely high. This is because public laboratories have a great deal of agricultural research, in which personnel expenses are relatively high, and semi-government laboratories include a number involved in nuclear energy and space development research, requiring large facilities and equipment.

Fig. 3-30. Research Expenditures by Speciality

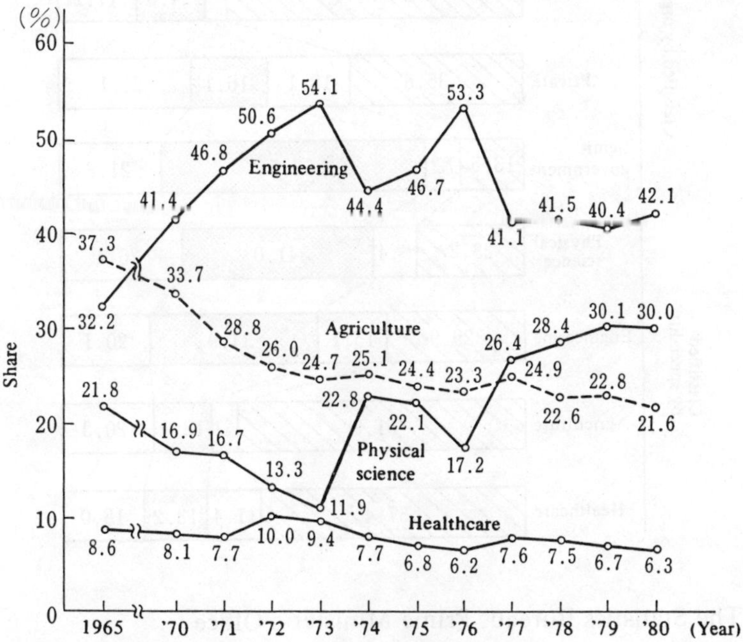

Source: The Statistics Bureau, Prime Minister's Office

Fig. 3-31. Research Expenditures at Research Laboratories by Expense

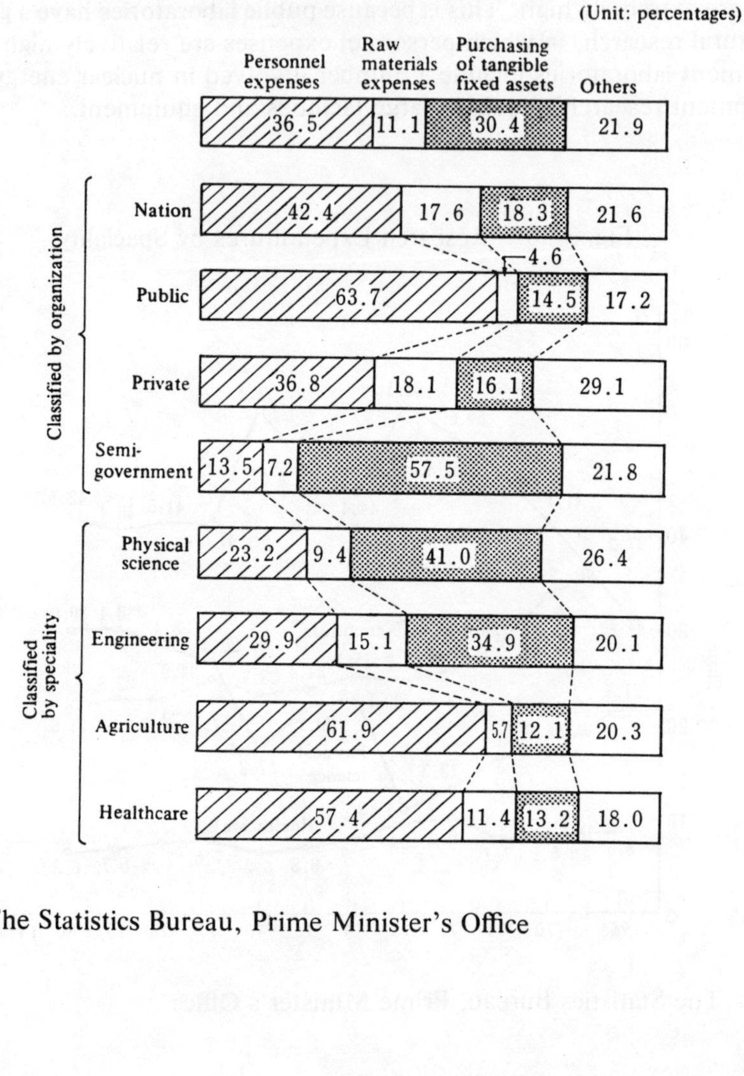

Source: The Statistics Bureau, Prime Minister's Office

As shown in Figure 3-32, at national laboratories, the proportion of basic research is large, while at public and private laboratories the proportion of applied research is large. Semi-government laboratories are characterized by the fact that the proportion of developmental research is conspicuously high.

Fig. 3-32. Research Expenditures at Research Laboratories by Nature of Research

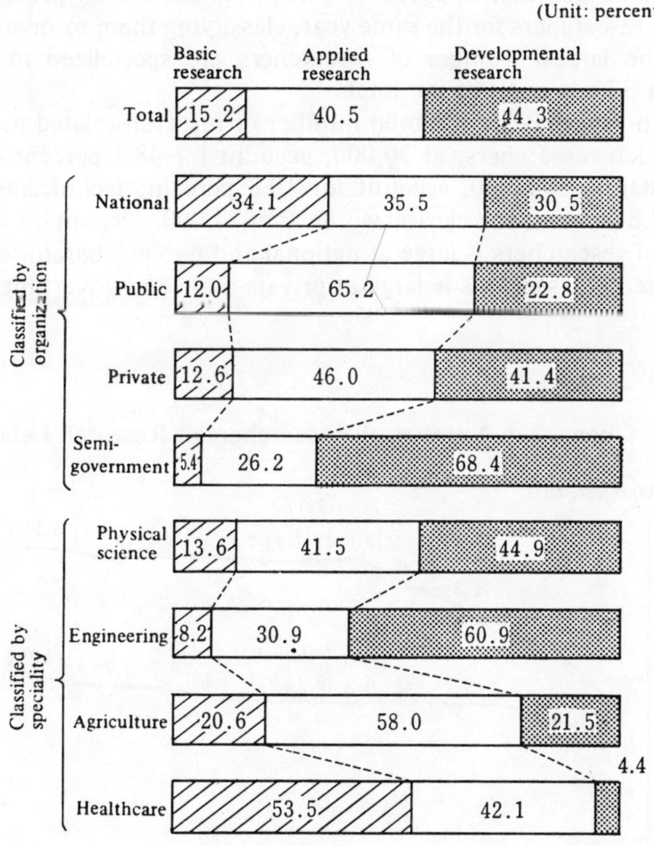

Source: The Statistics Bureau, Prime Minister's Office

For research laboratories as a whole, research expenditures per researcher came to $100,240 in 1980, a 12.6 percent increase over the previous year. By type of organization, the figure of $434,080 for semi-government laboratories is the highest, followed by $195,120 for private laboratories, $74,920 for national laboratories, and $47,440 for public laboratories.

As shown in Figure 3-33, the number of researchers at research laboratories in 1981 was 30,000, a 4.8 percent increase over the previous year. National laboratories accounted for 10,100 (33.6 percent of the total), public laboratories accounted for 14,100 (47.0 percent), laboratories accounted for 3,400 (11.4 percent), and semi-government laboratories 2,400 (8 percent). Table 3-34 presents a breakdown of these researchers for the same year, classifying them by organization and speciality. The largest number of researchers are specialized in agriculture, accounting for 37.6 percent of the total.

As shown in Figure 3-35, the total number of research-related personnel was 61,000, of which researchers, at 30,000, account for 48.8 percent of the total; research assistants, at 6,000, account for 10.8 percent; technicians, at 11,000 account for 17.8 percent; and clerical workers, at 14,000, account for 22.5 percent. The number of researchers is large at national and public laboratories, while the number of research assistants is large at private and semi-government laboratories.

Fig. 3-33. Changes in Number of Researchers at Research Laboratories

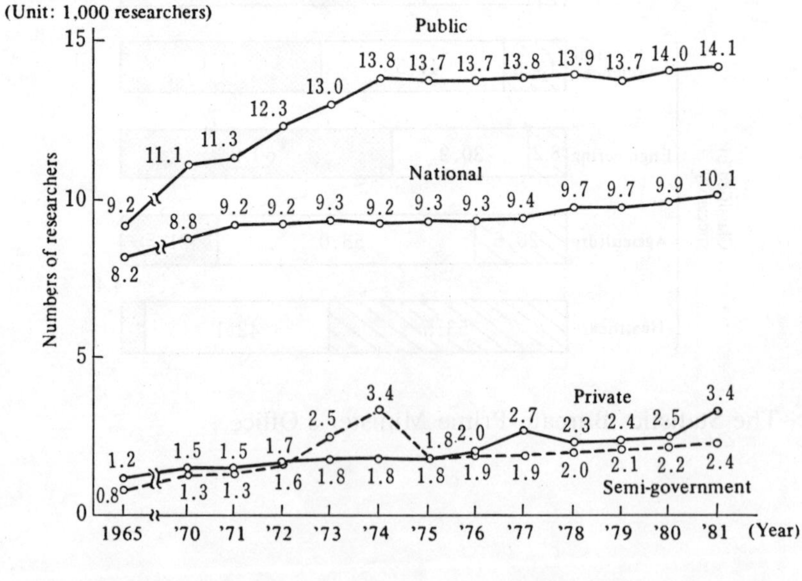

Source: The Statistics Bureau, Prime Minister's Office

Table 3-34. Number of Researchers at Research Laboratories Classified by Organization and Speciality (1981)

(Unit: number of researchers)

Classified by speciality / Classified by organization	Total numbers	Physical science	Engineering	Agriculture	Healthcare
Total number (proportion in percentage)	30,006 (100.0)	5,294 (17.6)	10,240 (34.1)	11,287 (37.6)	3,185 (10.6)
National	10,073	1,444	4,317	3,196	1,116
Public	14,110	1,589	3,312	7,735	1,474
Private	3,412	379	2,139	299	595
Semi-government	2,411	1,882	472	57	–

Source: The Statistics Bureau, Prime Minister's Office

Fig. 3-35. Research-Related Personnel at Research laboratories

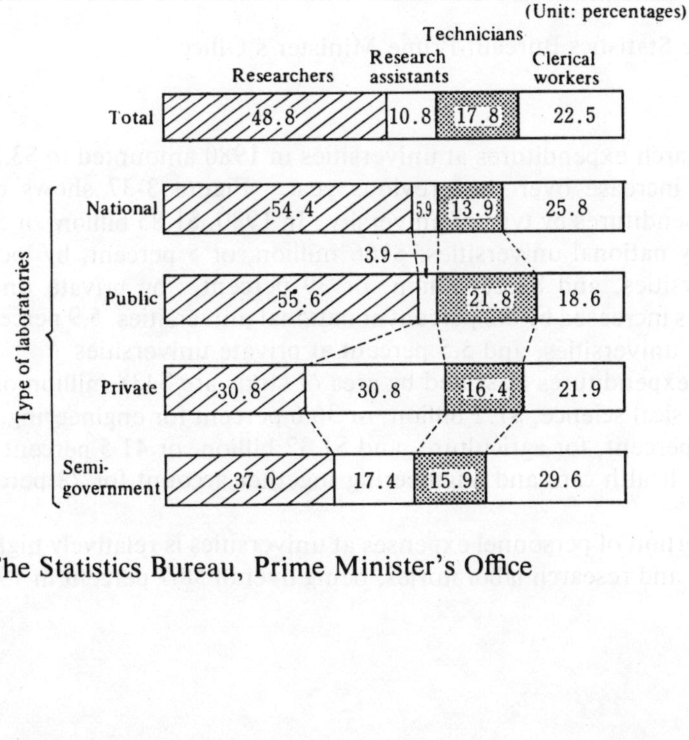

Source: The Statistics Bureau, Prime Minister's Office

4. UNIVERSITIES

Universities, responsible for training future research personnel, also fulfill an extremely important role in advancing basic research.

As shown in Table 3-36, the number of university departments, in Japan in 1980 amounted to 643, a 1.6 percent increase over the previous year. National universities account for 56.8 percent, local government universities for 7.0 percent, and private universities for 36.2 percent.

Table 3-36. Changes in Number of University Departments

Classification \ Year	1965	1970	1975	1976	1977	1978	1979	1980
National	264	267	318	334	339	349	359	365
Local government	40	38	39	40	41	42	44	45
Private	143	179	221	227	224	227	230	233
Physical science	64	70	80	81	85	85	86	84
Engineering	209	222	249	261	260	263	268	279
Agriculture	61	64	71	71	69	71	70	71
Healthcare	113	128	178	188	190	199	209	209
Total	447	484	578	601	604	618	633	643

Source: The Statistics Bureau, Prime Minister's Office

Total research expenditures at universities in 1980 amounted to $3.3 billion, a 5.9 percent increase over the previous year. Figure 3-37 shows changes in research expenditures by type of university. In 1980, $1.85 billion, or 56 percent, was spent by national universities, $166 million, or 5 percent, by local government universities, and $1.28 billion, or 39 percent, by private universities. Expenditures increased by 6.2 percent at national universities, 5.9 percent at local-government universities, and 5.5 percent at private universities.

Research expenditures classified by area of study are $438 million or 13.3 percent, for physical science, $1.2 billion, or 36.6 percent for engineering, $284 million, or 8.6 percent, for agriculture, and $1.37 billion, or 41.5 percent for health care, so that health care and engineering together account for 78 percent of the total.

The proportion of personnel expenses at universities is relatively higher than at corporations and research laboratories, being overall 58.7 percent in 1980, and at

Fig. 3-37. Changes in Research Expenditures at Universities

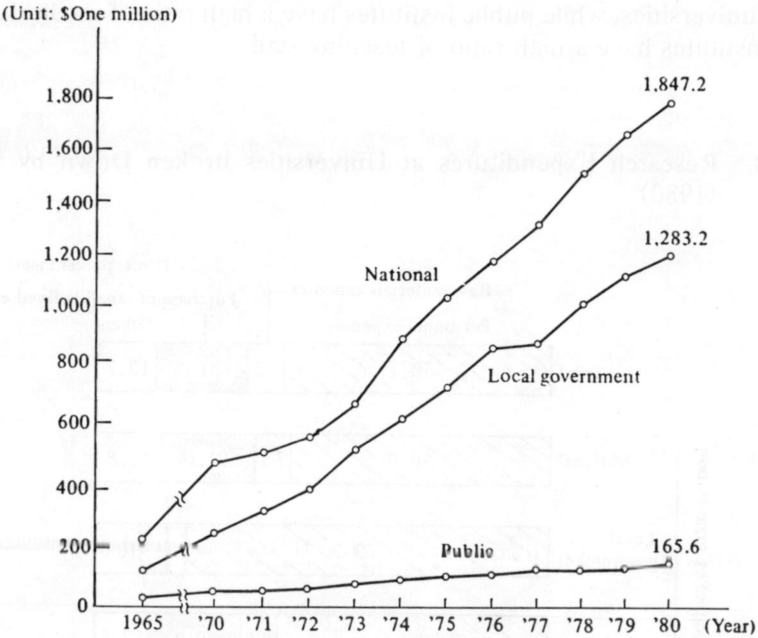

Source: The Statistics Bureau, Prime Minister's Office

local government universtities alone as high as 79.9 percent, as shown in Figure 3-38.

Not surprisingly, for universities the share of expenditures for basic research exceeds 50 percent, while that for developmental research is relatively low, as shown in Figure 3-39.

Research expenditures per researcher at universities as a whole increased by 1.7 percent from $32,160 in 1979 to $32,720 in 1980. The figure for teaching staff alone was $48,640, which, broken down by type of university, comes to $50,800 for national, $47,760 for local government, and $36,800 for private universities.

The number of researchers at universities in 1981 was 103,000, a 1.9 percent increase over the previous year. As seen in Figure 3-40, the number of researchers increased 0.2 percent over the previous year at national universities, 0.7 percent at local government universities, and 4.9 percent at private universities.

The number of research-related personnel at universities was 137,000, of which researchers, at 103,000, account for 74.7 percent of the total, as shown in Figure 3-41. Research assistants amounted to 8,000, or 5.9 percent of the total; technicians, 12,000, or 8.8 percent; and clerical workers, 15,000, or 10.6 percent.

Researchers at universities can be classified into teaching staff, graduate school students, and medical staff. The ratio of graduate school students is high at national universities, while public institutes have a high ratio of medical staff and private institutes have a high ratio of teaching staff.

Fig. 3-38. Research Expenditures at Universities Broken Down by Expense (1980)

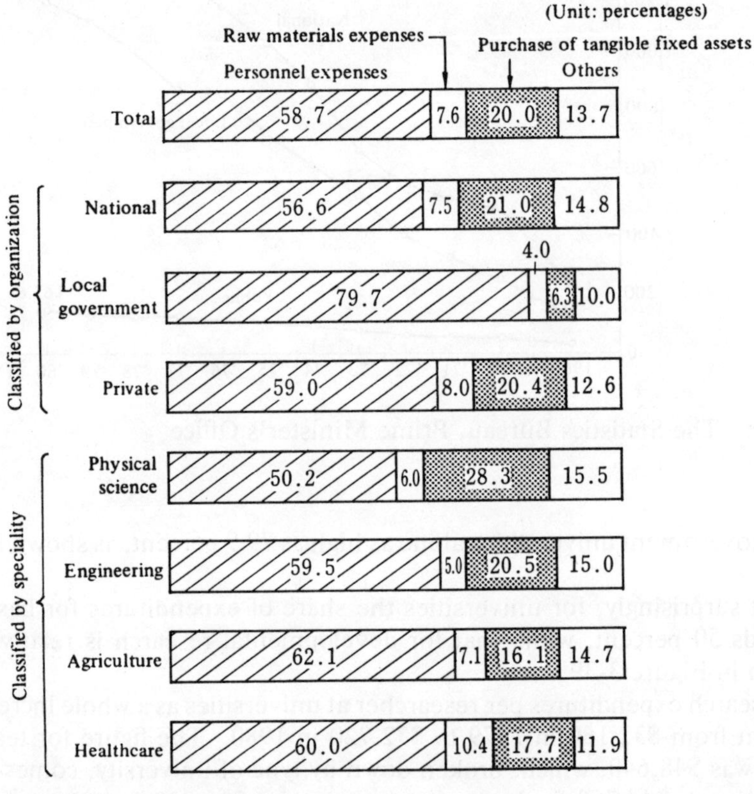

Source: The Statistics Bureau, Prime Minister's Office

Fig. 3-39. Research Expenditures at Universities Broken Down by Nature of Research

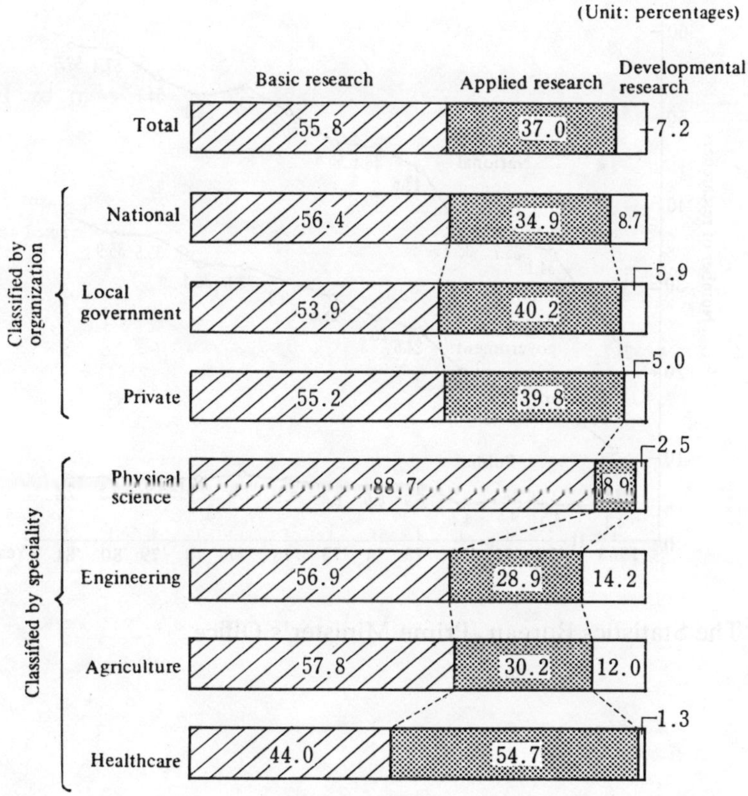

Source: The Statistics Bureau, Prime Minister's Office

Fig. 3-40. Changes in Number of Researchers at Universities

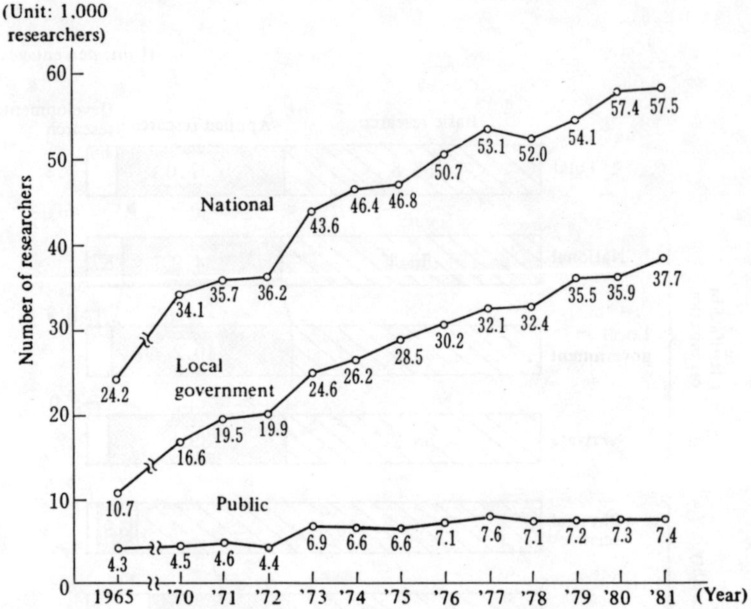

Source: The Statistics Bureau, Prime Minister's Office

Fig. 3-41. Ratio of Research-Related Personnel at Universities

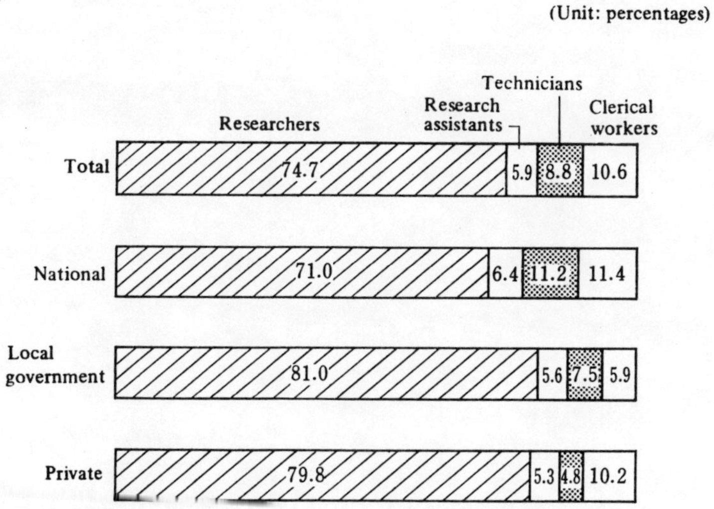

Source: The Statistics Bureau, Prime Minister's Office

CHAPTER 4

COLLECTION AND DISTRIBUTION OF SCIENTIFIC INFORMATION

Much of the knowledge acquired in the course of scientific research is passed on to the public as various forms of information. This information, the result of activities in which large amounts of time, money and labor have been invested, is then used to develop new knowledge.

With the growth of science and technology, the volume of information produced is doubling every 10 to 15 years. Counting only public documents such as journal articles, patent information, and technical reports, it is estimated that the amount exceeds approximately 4.5 million items annually. Because of this flood of information, it has become increasingly difficult to obtain quick access to necessary information. As a result, great losses are caused by unnecessary duplication of successful research, or by needless repetition of unsuccessful experiments, due to lack of appropriate information.

Under these circumstances, the importance of storage and retrieval of information is constantly increasing. In particular, with the spectacular growth in computers and data communications in recent years, public and private institutions around the world have become active in setting up data bases.

Scientific information services in Japan are being carried out by a variety of institutions, such as those listed in Table 4-1. The majority of these institutions collect documents, abstract or index them, and make them available to users. Some go a step further by offering mechanical retrieval services.

Table 4-1. Major Public Scientific Information Services in Japan

Field	Institution	Nature of activities
Science and technology in general	Japan Information Center of Science and Technology (JICST)	o Implements NIST activities o Gathers and distributes scientific information o Serves as a clearing house for information on all public research laboratories and some private research laboratories in Japan
Patents	Japan Patent Information Center Co., Ltd. (JAPATIC)	o Provides bibliographic and descriptive information on patents and utility model registrations o Provides bibliographic information on U.S. patents o Publishes abstracts of patent information in cooperation with the Japan Institute of Invention and Inovation o Exchanges patent information with the International Patent Documentation Center
Agriculture, forestries and fisheries	Agriculture, Forestries, and Fisheries Research Information Center	o Gathers and indexes domestic literature on agriculture, forestry, and fishery o Operates a mechanical retrieval system for research subjects of public laboratories in the field of agriculture, forestry, and fisheries. o Provides information to AGRIS, an international agricultural information system, and the Food and Agriculture Organization of the United Nations
Medicine	International Medical Information Center	o Gathers and distributes domestic and foreign medical information o Provides medical and biological information to the International Nuclear Information System (INIS) o Provides information on cancer to the International Cancer Research Data Base (ICRDB).
Healthcare	Health and Welfare Statistics Association Japan Pharmaceutical Information Center	o Publishes vital statistics and hygiene statistics o Provides abstracts and indexes of literature on pharmaceutical products
Biology	Institute of Physical and Chemical Research	o Gathers information on experimental life forms, microbes, seaweeds, animal and plant cells, and plant life
Chemistry	The Japan Association for International Chemical Information, Japan Phamaceutical Information Center	o Publishes bulletins on domestic chemical information o Cooperates with Chemical Abstracts Service (CAS) of the U.S.
Metals	The Iron and Steel Institute of Japan (The Japan Iron and Steel Technical Information Center)	o Gathers information on iron, steel, and other metals.
Labor Safety	Mining Labor Accident Prevention Association	o Gathers information on labor safety and hygiene o Provides information to CIS, an international labor safety information center
Environmental pollution	National Institute for Environmental Pollution Research	o Gathers information on environmental pollution o Belongs to the International Referral System for Sources of Environmental Information (INFOTERRA)
Disaster prevention	National Research Center for Disaster Prevention, Earthquake Research Institute, University of Tokyo	o Gathers information on disaster prevention science o Gathers data on earthquakes
Nuclear energy	Japan Atomic Energy Research Institute	o Gathers information on nuclear energy o Provides information to the International Nuclear Energy Information System (INIS)
Oceanography	Japan Marine Science and Technology Center	o Offers information on marine development technology
Medium and small businesses	The Small Business Corporation (The Small Business Information Center)	o Provides source and analysis of technological information for medium and small businesses

The Japan Information Center of Science and Technology (JICST), the key center for scientific and technological information in Japan, as of 1971 had stored in its data base some 450,000 documents gathered from all over the world. In 1976, it began the JICST On-line Information System (JOIS), which in 1981 was upgraded to the JOIS-II, comparable to systems in Europe and the United States. The Ministry of Education, utilizing a network of seven large computer centers such as those at Tokyo University and Kyoto University, offers an on-line service for universities. Also, in light of the importance of creating a more effective information distribution system at higher-level academic and research institutions, the Ministry of Education is studying the feasibility of an information center and a data base system.

In 1980, the Agency of Industrial Science and Technology, part of the Ministry of International Trade and Industry, established the Research Information Processing System (RIPS), linking research laboratories with optical fibers.

There are a number of centers that gather numerical data obtained through tests, measurements, and observations, then analyze and evaluate the information, such as the Tokyo Astronomical Observatory (natural sciences), the Institute for Protein Research (bio-medicine), the Japan Atomic Energy Research Institute (physics), and the National Research Institute for Metals (engineering). Some of these centers have accumulated a considerable amount of data, but unfortunately most of them belong to the government and the data is not open to the public.

In physics and engineering, because of public demand, the number of private data centers is increasing, but the amount of data accumulated so far is still relatively small.

As far as libraries are concerned, the National Diet Library is the largest in Japan, and every work published in Japan is delivered to it. In addition to weekly bulletins on newly received books, the National Diet Library also offers automatic editing of listings and indexes for easy retrieval of materials in its collection. The number of scientific and technical journals subscribed to amounts to approximately 4,500 from Japan and 8,500 from abroad.

The Ministry of Education publishes a general listing of scientific journals subscribed to by libraries of universities and major research laboratories. The natural science section of this listing contains approximately 36,800 European and American journals and 25,000 Japanese journals.

Information service that goes a step further by analyzing the information stored and making it available in response to requests from users is offered by many institutions. But the sophisticated techniques and specialized knowledge required for its use make it impossible to fulfill current needs. Enterprises that view such service with particular importance include the Small Business Corporation, which is providing technological and management information to regional public bodies, and the International Medical Information Center, which provides user-requested studies of medical information as one of its main activities. As a clearing house for information on researchers and research subjects, the Japan Information

Center of Science and Technology provides an on-line information system concerning research in progress at public research laboratories in Japan.

Another important activity in scientific information services is the international cooperation being carried out both at governmental and private levels. At the governmental level, activities which include exchange of information and personnel are being conducted between Japan and such countries as the United States, West Germany, and France.

CHAPTER 5

TECHNOLOGY TRANSFER AND PATENT APPLICATIONS

International transactions in granting franchises and approving use of patents, utility models, and technical know-how are known as technological transfer, and together with applications for patents reflect the level of a country's research and development.

1. Technological Transfer

According to the Bank of Japan's monthly review of the country's international balance of trade, the value of technological exports from Japan in 1981 amounted to $473.6 million, an increase of 38 percent over the previous year, while imports came to $1.51 billion, an increase of 16 percent. The ratio of exports to imports was 0.31, an increase of 0.05 over the previous year, as shown in Figure 5-1. However, according to the Statistics Bureau of the Prime Minister's Office, technological trade exports in 1980 reached $638.4 million, an increase of 19.9 percent from the previous year, and imports amounted to $958 million, a decrease of 0.6 percent. For newly contracted portions only, total technological exports exceeded imports from 1972 onwards, and in 1980 the figures were $297.2 million and $110.8 million, respectively.

Fig. 5-1. Changes in Amount of Technological Trade in Japan

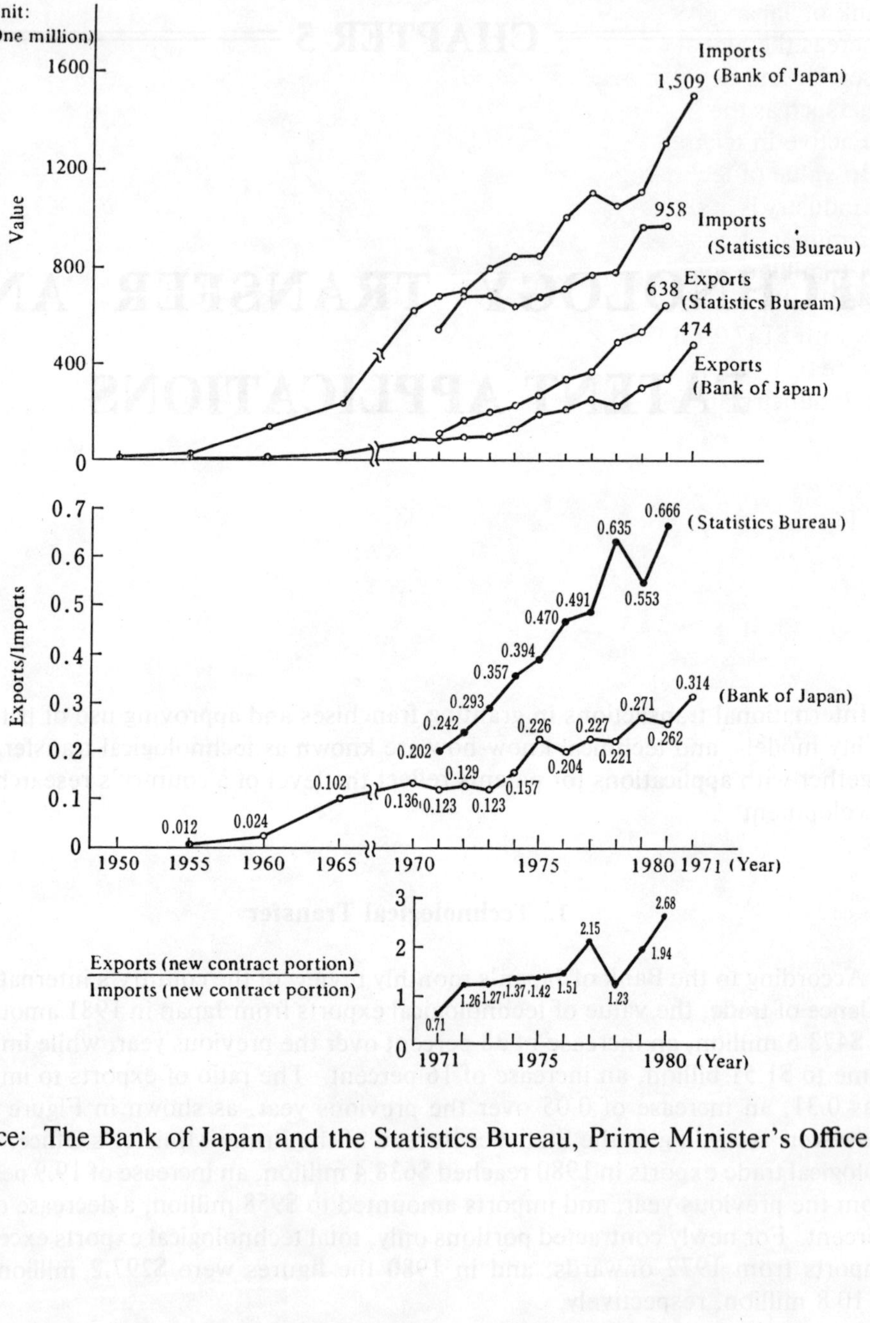

Source: The Bank of Japan and the Statistics Bureau, Prime Minister's Office

The Statistics Bureau's larger figures for exports cover all technological trade, while the Bank of Japan's cover only technical assistance. On the other hand, the Bank of Japan gives a larger figure for imports, because it covers all industries, whereas the Statistics Bureau limits its figures to companies conducting domestic research and development, thus excluding semi-government research laboratories (such as the National Space Development Agency) and service industries that are active in technological trade.

In value of technological exports, among manufacturing industries the chemical industry is at the top with $127.6 million, followed by electrical machinery with $92 million, transportation machinery with $87.2 million, and iron and steel with $71.6 million. For imports, the electrical machinery industry is at the top with $246.8 million, followed by transportation machinery with $161.2 million, chemicals with $157.2 million, and general machinery with $120.8 million, as shown in Figure 5-2. Main industries with exports greater than imports are the iron and steel industry and the textile industry.

Fig. 5-2. Breakdown of Japan's Technological Trade by Industry (1980)

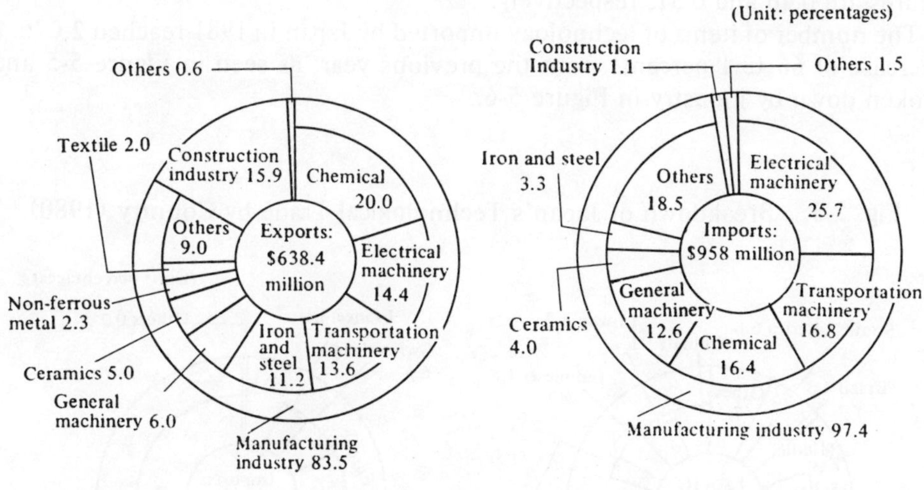

Source: The Statistics Bureau, Prime Minister's Office

As shown in Figures 5-3 and 5-4, a large portion of Japan's technological exports is directed to Southeast Asian countries, which amounted to a total of $216.8 million in 1980, a decrease of 1.1 percent from the previous year. Of the Southeast Asian countries, China accounts for $77.6 million (including $39.6 million for Taiwan), Indonesia, $30.4 million, and South Korea, $21.2 million. Exports to the United States amounted to $88.4 million, an increase of 15.1 percent. Technology imports from the United States are substantial every year, and amounted to $615.2 million in 1980, almost the same as the previous year. This was followed by imports from West Germany, $82 million, a decrease of 6.4 percent, and from England, $80.8 million, an increase of 16.1 percent.

Among leading countries, technical exports from the United States were by far the largest at about $7.1 billion (figure for 1981), far outdistancing France at about $1.8 billion (1977), England at about $857 million (1979), West Germany at about $495 million (1979), and Japan at about $473 million (1981). As opposed to this, in terms of technological imports, Japan is at the top with about $1.5 billion (1981), followed by France with about $1.3 billion (1971), West Germany with $1.1 billion (1979), the United States with $883 million (1981), and England with $665 million (1979). In terms of the ratio of exports to imports, the United States is number one at 8.03, followed by France at 1.43 and England at 1.29. In the case of West Germany and Japan, where imports exceeds exports, the figures are 0.46 and 0.31, respectively.

The number of items of technology imported by Japan in 1981 reached 2,076, a decrease of 66 (3.1 percent) from the previous year, as seen in Figure 5-5 and broken down by industry in Figure 5-6.

Fig. 5-3. Breakdown of Japan's Technological Trade by Country (1980)

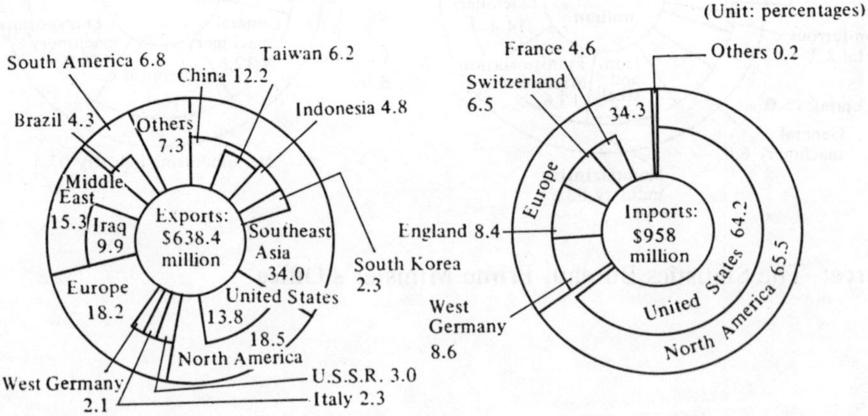

Source: The Statistics Bureau, Prime Minister's Office

Fig. 5-4. Breakdown of Japan's Technological Trade by Region and Industry (1980)

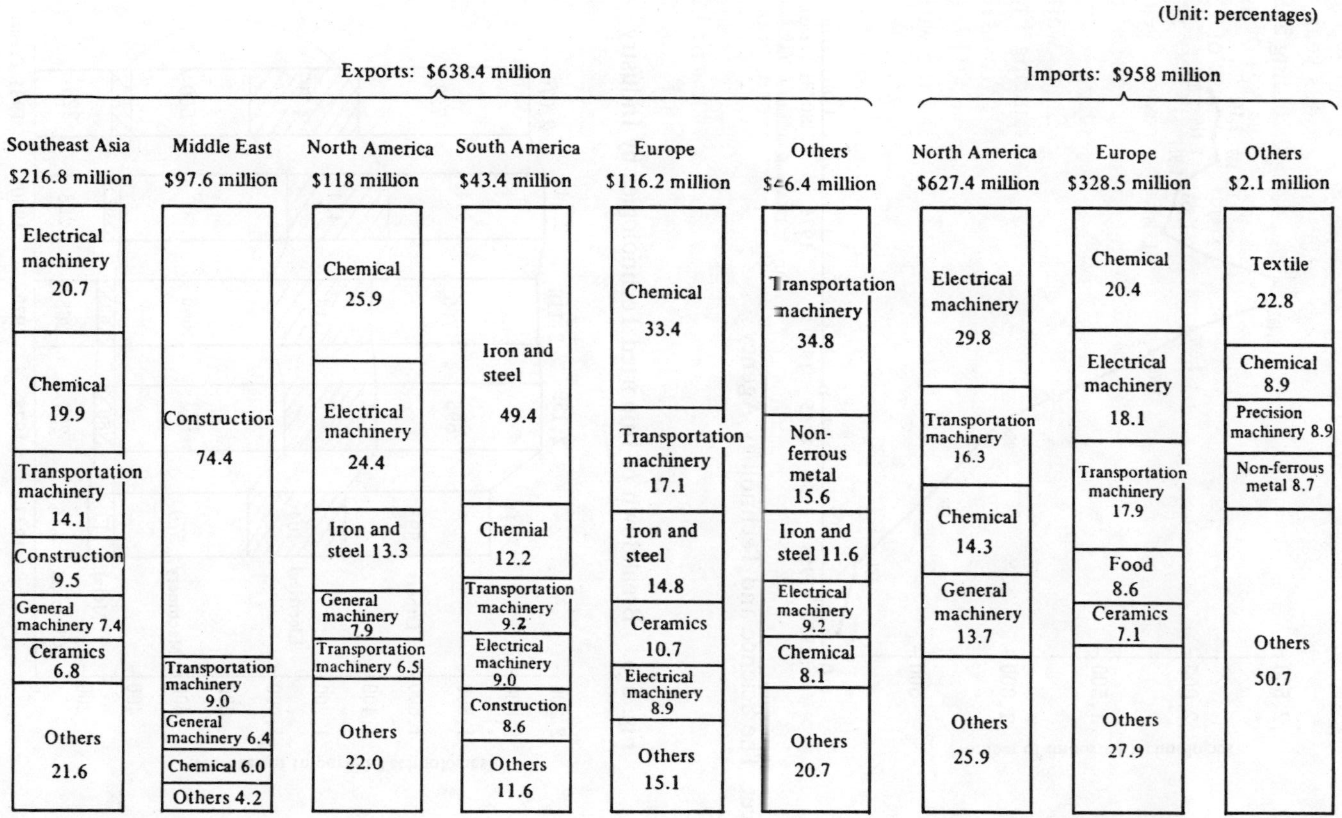

Source: The Statistics Bureau, Prime Minister's Office

Fig. 5-5. Changes in Number of Technologies Imported

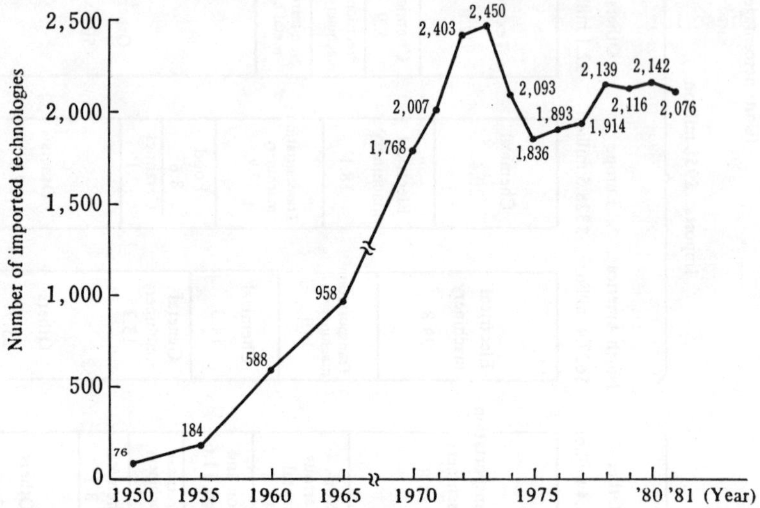

Source: The Science and Technology Agency

Fig. 5-6. Breakdown of Imported Technologies by Industry

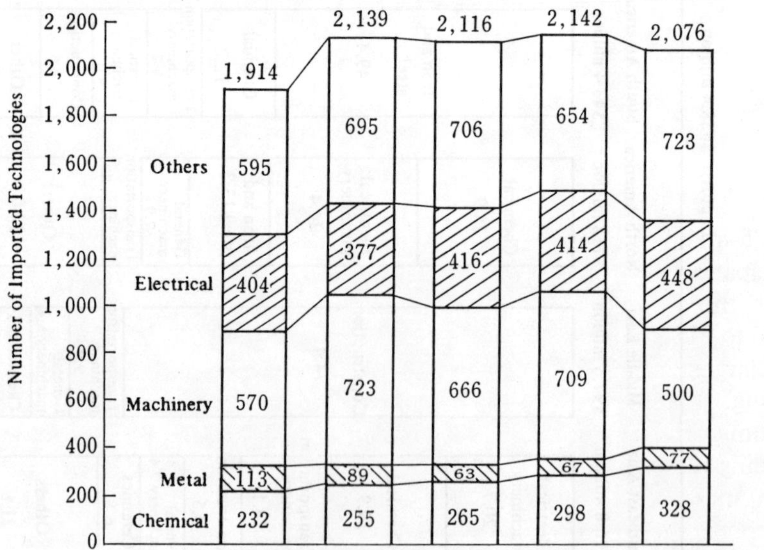

Source: The Science and Technology Agency

As seen in Figure 5-7, by far the largest number of technologies were from the United States, totalling 977 (47.1 percent), followed by France with 224 (10.8 percent), West Germany with 219 (10.5 percent), and England with 175 (8.4 percent). These four countries alone accounted for 76.8 percent of the total.

Fig. 5-7. Main Countries Exporting Technology to Japan

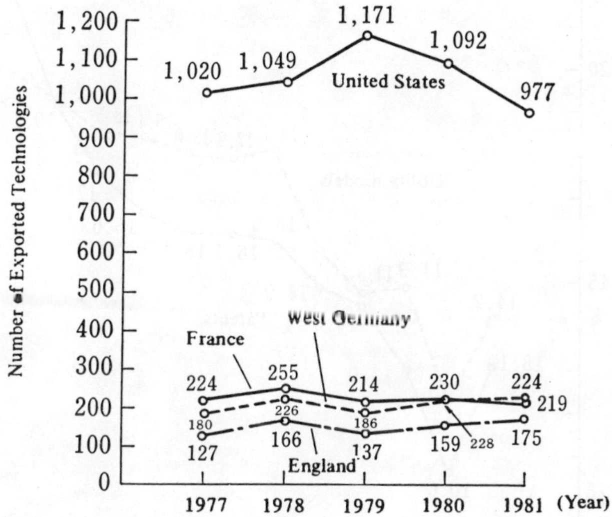

Source: The Science and Technology Agency

2. Patent Applications

Figure 5-8 shows changes in the number of patent and utility model applications in Japan. The sharp drop in 1971 was due to revision of the patent laws, but thereafter the number continued to grow, and in 1981 patent applications amounted to 218,261 and utility model applications, 198,978. As shown in Figure 5-9, the largest number of patents applied for were in the field of electrical engineering, totalling 41,693 (22.0 percent), followed by those in physics with 41,123 items (21.7 percent), and disposition, operations, and transportation with 37,711 items (20.0 percent). Compared with the previous year, there was about a 10 percent increase in the total number of items, but there was hardly any change in the composition.

Fig. 5-8. Changes in Number of Patent and Utility Model Applications

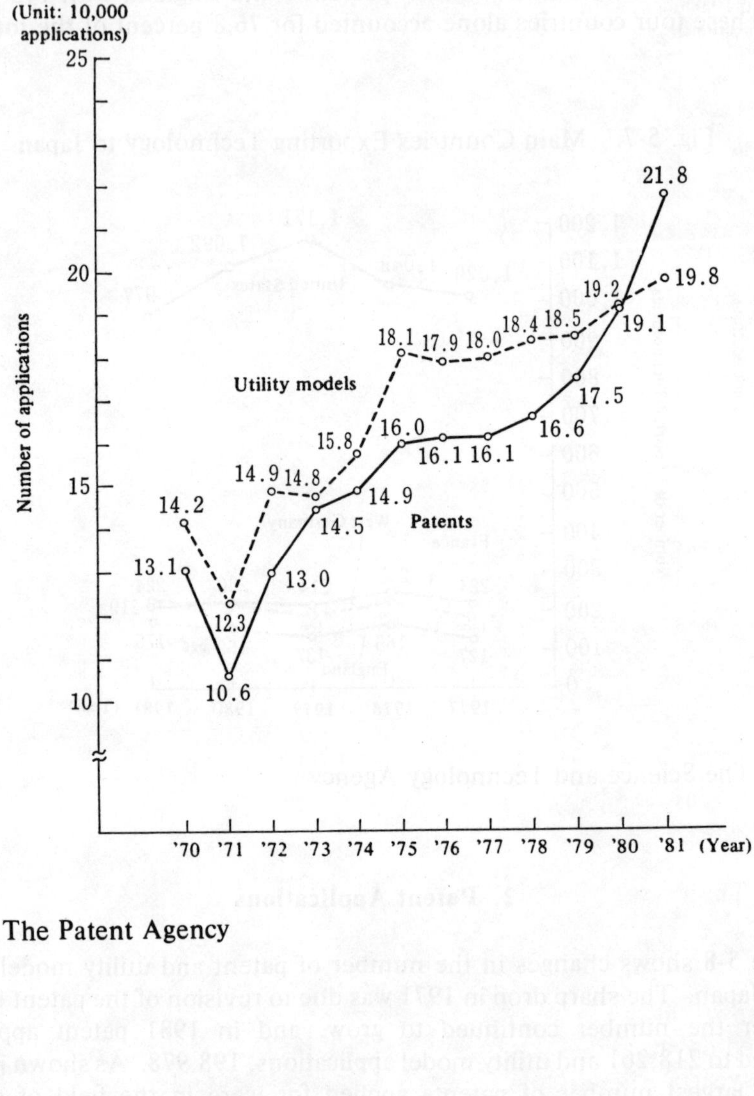

Source: The Patent Agency

Fig. 5-9. Changes in the Number of Patent Applications by Category

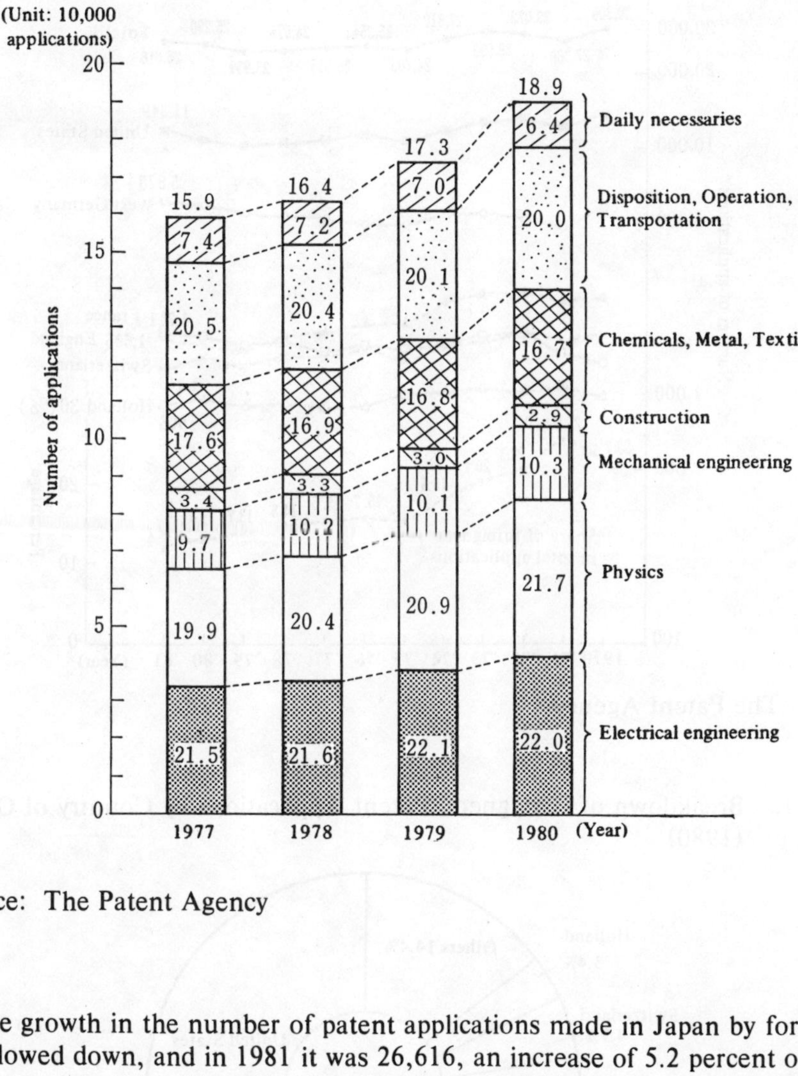

Source: The Patent Agency

The growth in the number of patent applications made in Japan by foreigners has slowed down, and in 1981 it was 26,616, an increase of 5.2 percent over the previous year. As shown in Figures 5-10 and 5-11, by country of origin the United States was the largest with 42.6 percent, followed by West Germany with 22.1 percent. Table 5-12 shows that the largest number of applications by foreigners was made in the category of chemicals, metals, and textiles.

Fig. 5-10. Changes in the Number of Patent Applications by Foreigners to Japan

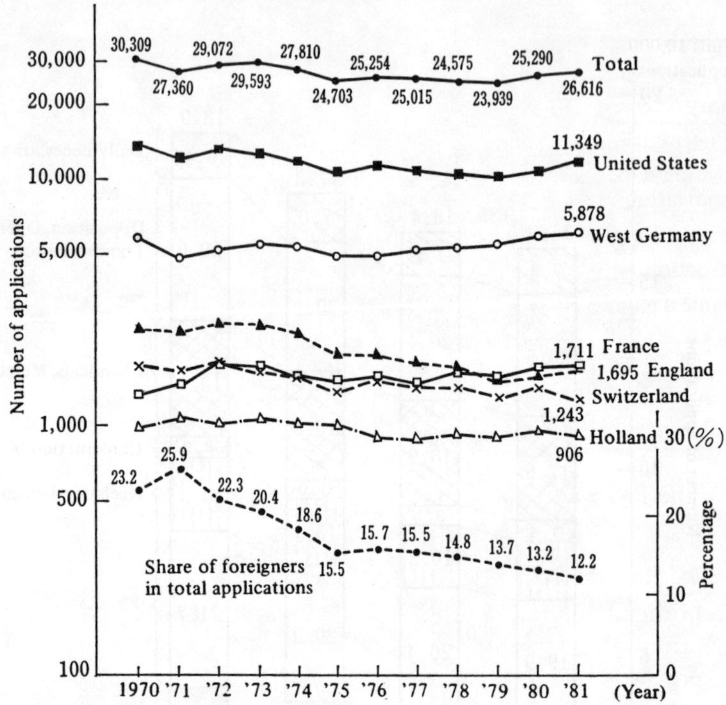

Source: The Patent Agency

Fig. 5-11. Breakdown of Foreigners' Patent Applications by Country of Origin (1980)

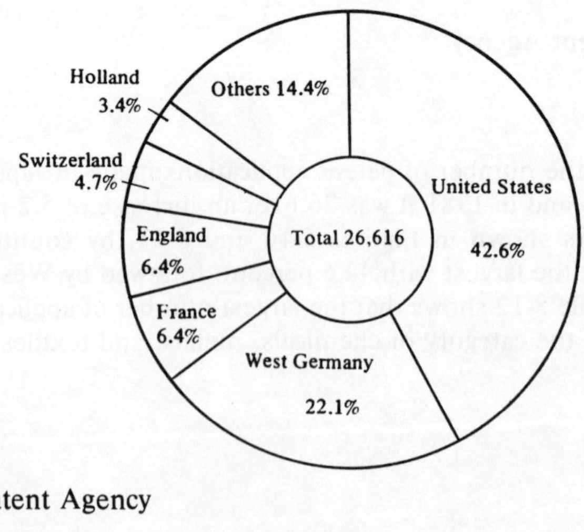

Source: The Patent Agency

Table 5-12. Breakdown of Foreigners' Patent Applications by Category

(Unit: percentages)

Category \ Year	1976	1977	1978	1979	1980
Daily necessaries	16.4	16.5	15.9	15.6	16.1
Disposition, Operation, Transportation	15.5	16.1	15.5	14.5	13.3
Chemicals, Metal, Textiles	26.8	26.2	25.6	23.9	23.7
Construction	10.9	10.7	11.1	10.9	10.7
Mechanical engineering	18.6	19.1	17.3	16.9	15.3
Physics	11.6	11.7	11.0	10.2	9.2
Electrical engineering	9.2	9.1	8.8	8.5	8.2
Total	15.7	15.6	14.9	14.1	13.3

Source: The Patent Agency

On the other hand, the number of patent applications by Japanese to foreign countries amounted to 35,254, as seen in Figure 5-13, an increase of 11.9 percent over the previous year. Figure 5-14 shows that the largest number of applications was made to the United States, amounting to 36.7 percent, or about one-third of the total, followed by West Germany with 14.9 percent, England with 12.0 percent, and France with 7.1 percent. Furthermore, in leading industrial countries, the share in the number of patent applications by Japanese is increasing. As shown in Table 5-15, the share of Japanese-held patents in 1980 is four or five times as large as it was in 1965.

Fig. 5-13. Changes in the Number of Patent Applications by Japanese to Major Countries

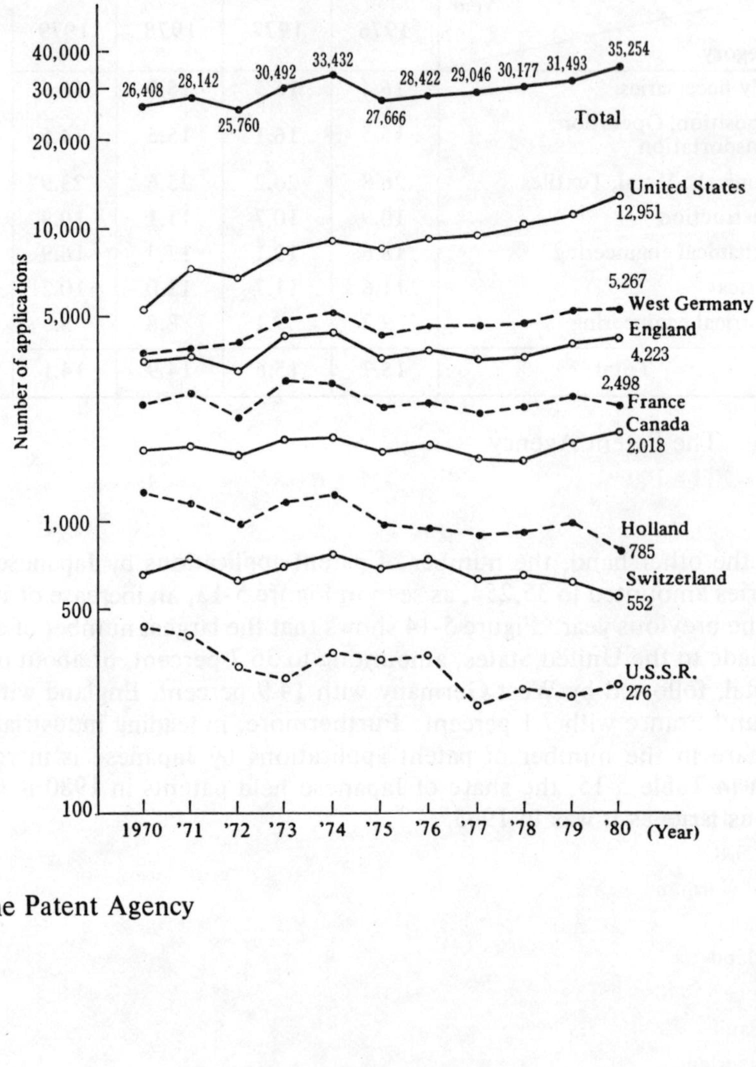

Source: The Patent Agency

Fig. 5-14. Breakdown of Japanese Patent Applications by Country (1980)

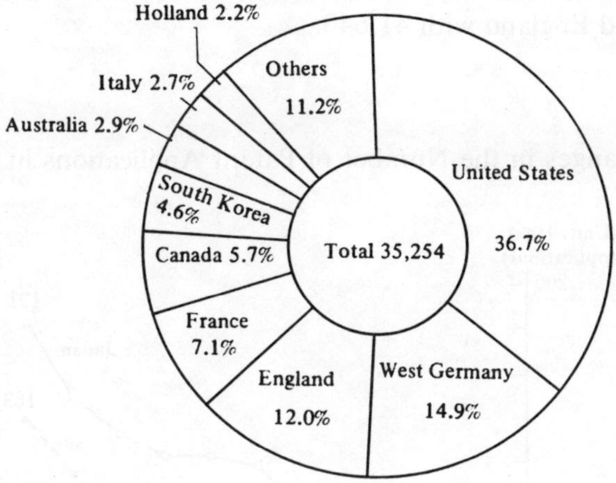

Source: The Patent Agency

Table 5-15. Changes in the Share of Japanese Patent Applications in Major Countries

(Unit: percentages)

Year / Country	1965	1970	1975	1976	1977	1978	1979	1980
United States	2.4	5.1	8.5	9.2	9.6	10.1	11.1	12.4
U.S.S.R.	–	8.8	7.2	8.2	6.2	8.0	7.4	8.8
West Germany	2.0	5.8	7.2	7.5	7.7	8.1	9.6	10.8
Canada	1.6	5.8	6.8	7.0	12.9	6.5	7.8	8.1
England	2.3	5.8	6.8	7.0	6.6	7.3	9.1	10.1
France	2.0	5.3	6.0	6.4	5.9	6.6	8.3	8.9
Holland	1.6	6.5	6.4	6.4	6.1	7.1	10.5	10.7
Switzerland	1.4	3.4	4.1	4.5	3.9	5.0	5.3	5.7

Source: The Patent Agency

Next, Figure 5-16, giving the number of patent applications in leading countries, shows that in 1980 Japan was at the top with 191,020 applications, followed by the Soviet Union with 168,589, the United States with 104,329, West Germany with 48,583, and England with 41,640.

Fig. 5-16. Changes in the Number of Patent Applications in Major Countries

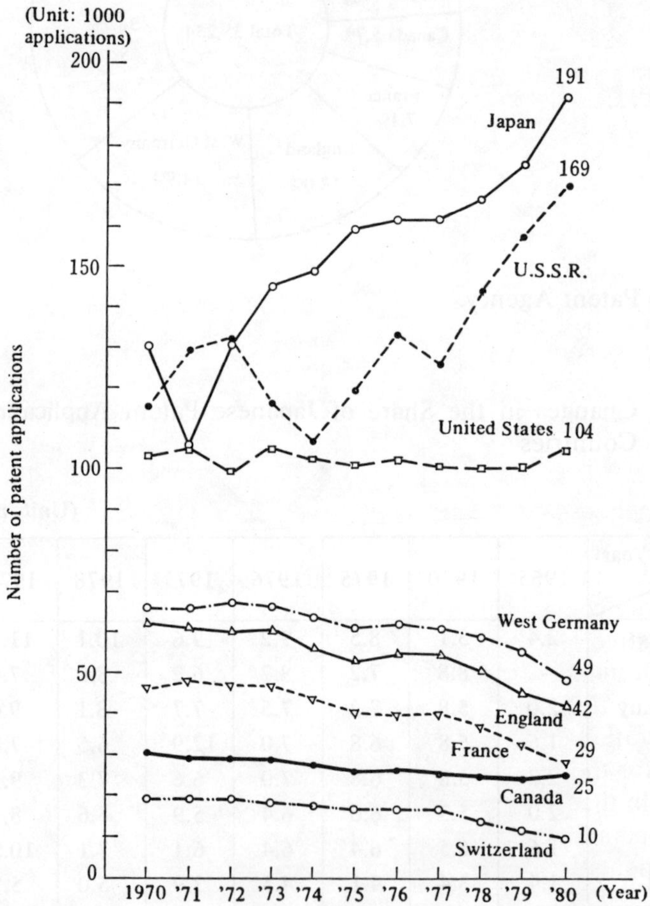

Source: The Patent Agency

CHAPTER 6

INTERNATIONAL COOPERATION

International cooperation in scientific and technological research projects too large for one country to handle is being increasingly pursued through government, private, and international organizations, and ranges from simple exchange of data to more complex, full-scale joint research. In addition, the objectives of these cooperative efforts can be both scientific and political.

Japan's international cooperation in science and technology got underway in the first half of the 1950's, when the nation's industry recovered to its pre-war level. Cooperation during this period involved mostly exchange of personnel and information for the purpose of assimilating knowledge from industrialized countries. Then, in the first half of the 1960's, a stronger foundation for international scientific cooperation was gradually laid. As can be seen in Figure 6-1, the number of agreements in science and technology has continued to increase, especially in the early 1970's, when the level of technology in Japan rose rapidly.

Fig. 6-1. Aggregate Number of International Agreements Concerning Scientific Cooperation with Japan

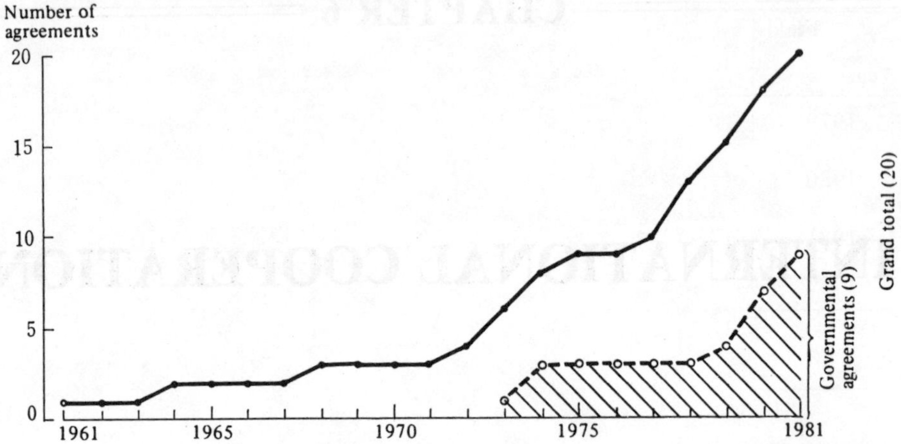

Note: Agreements include only those involving public research laboratories and universities, but excluding agreements concerning cooperation in just one specific field.

Source: The Science and Technology Agency

1. OUTLINE OF INTERNATIONAL AGREEMENTS

Japan is cooperating with other nations in various areas, including atomic energy, nuclear fusion, life science, and space technology. A summary of these activities by country will be given in this section.

(1) The United States

The United States, both politically and economically one of the closest nations to Japan, is also the most advanced nation in the world in science and technology. As a result, Japan has more agreements for scientific cooperation with the United States than with any other country.

Ever since the establishment of the Japan-U.S. Committee on Scientific Cooperation in 1961, there have been various agreements, such as those on natural resources (1964), environmental protection (1975), energy (1979), and non-energy fields (1980). The agreements between Japan and the United States in energy and non-energy fields have the largest budget among all agreements for international scientific cooperation in Japan. A breakdown of budgets for these two fields is shown in Table 6-2.

Table 6-2. Budgets for the Japan-U.S. Agreements in Energy and Non-energy Areas

(Unit: $one thousand)

Field / Year	Nuclear fusion	Coal liquefaction	Photo-synthesis	High energy physics	Geothermal energy	Total
1979	10,200	3,100 (1,800)	320	900	0	14,520 (13,200)
1980	15,000	36,200 (6,200)	940	5,500	2,300	59,940 (29,940)
1981	17,200	71,300 (10,800)	1,050	6,600	2,300	98,450 (37,950)
1982	14,400	8,200	804	6,600	2,348	32,352

(Unit: $one thousand)

Field / Year	Space	Nuclear physics	Life science	Others	Total
1980	340	0	0	0	340
1981	940	470	20	30	1,460
1982	1,860	920	20	30	2,830

Note: Figures in parentheses are amounts exclusive of expenditures for the SRC-II coal liquefaction project, which was cancelled in 1981.

Source: The Science and Technology Agency

Based on the agreement in the energy field, progress has been made in executing joint projects in nuclear fusion, coal liquefaction, photo-synthesis of solar energy, geothermal energy, and high-energy physics. Of these, development of nuclear fusion, because of the time and expenses involved, has from the beginning been regarded as one of the most important fields, and several projects, such as the Doublet III, are being conducted.

Joint projects based on the agreement in the non-energy field are being carried out in such areas as space technology, life science, nuclear physics, environmental protection, disaster prevention, and agriculture, in the form of exchange of information, conferences, and joint research projects. The first meeting of the Joint Committee based on this agreement was held in Tokyo in September, 1981, and discussed scientific and technological policies of both the Japanese and the United States governments and methods of furthering each project.

(2) France

Based on the Japan-France agreement on scientific co-operation concluded in July, 1974, joint research projects, conferences, and exchanges of information are being conducted in the field of marine technology. In the atomic energy field, since conclusion of the Japan - France agreement on atomic energy in 1972, active cooperation in peaceful uses of atomic energy has been carried out.

(3) West Germany

A scientific cooperation agreement with West Germany was concluded in October, 1974, on the basis of which joint projects are being carried out on oceanography, atomic reactor safety, high-temperature gas reactors, biology, medicine, environmental protection, new energy technology, space technology, and many other fields.

(4) Canada

An agreement with Canada for scientific cooperation was established in 1972, and to date four meetings have been held. Presently, besides cooperation in energy, space, agriculture, environment, and disaster prevention, joint studies are being conducted in design and paving of streets in cold weather districts, transportation of resources from northern regions, and other problems peculiar to high-latitude areas.

(5) Australia

The Japan-Australia agreement on science was concluded in November, 1980, under which cooperation in energy and other fields is advancing. Currently, cooperation in life science, plasma physics, earth science, and space technology is being planned.

(6) The U.S.S.R.

The Japan-Soviet agreement on scientific cooperation was concluded in October, 1973. Based on this agreement, cooperation in the fields of nuclear energy and agriculture is being conducted. Exchanges of researchers are also taking place in accordance with a cultural exchange agreement.

(7) East European Countries

The first agreement on scientific cooperation concluded with an East European country was made with Rumania in April, 1975, followed by East Germany in November, 1977, Bulgaria in March, 1978, Czechoslovakia and Poland in November, 1978, Hungary in May, 1979, and, most recently, Yugoslavia in May, 1981.

(8) China

An agreement between Japan and the People's Republic of China on cooperation in the field of science and technology was concluded in May, 1980. Based on this agreement, at a meeting held in Beijing in June, 1981, agreement was reached to conduct research on refining of iron ore containing niobium. Prior to this, technical cooperation in agriculture started in 1973, and, under the aegis of the Japan International Cooperation Agency, cooperation began in railroads in 1978, and business administration and medicine in 1979. Also, in accordance with a memorandum signed by the Japan Society for the Promotion of Science, the Chinese Ministry of Eduction, and the Chinese Academy of Science, exchange of researchers in scientific fields is being conducted.

(9) Indonesia

An agreement between Japan and the Republic of Indonesia on scientific cooperation was concluded in January, 1981. The first conference was held in Djakarta in January, 1982, during which agreement was reached on conducting research on use of plant activating organisms as agricultural chemicals.

(10) South Korea

As a result of talks begun in 1968, cooperaton is being extended in the fields of natural resources, energy, forestry, marine life, nuclear energy, environmental protection, and disaster prevention, among others.

2. TECHNOLOGICAL ASSISTANCE

Japan has contributed to improving the scientific and technological capabilities of developing countries in the form of training, guidance, and research. However, in comparison with achievements made by other leading countries, there is all the more need for Japan to cooperate more positively.

The total budget for bilateral scientific and technological assistance in 1980 was $252 million, an increase of $40 million, or 18.8 percent, over the 1979 figure of $212 million, as seen in Figure 6-3.

Among countries belonging to the Development Assistance Committee (DAC), the average ratio of bilateral assistance budgets to total government technology support funds was 20.4 percent for 1980. Japan's ratio, however, was only 8.4 percent, as shown in Figure 6-4. Further, Japan's share in the total bilateral cooperation funds of DAC countries was only 5.1 percent in 1980, significantly lower than France's 33.1 percent, West Germany's 18.1 percent, and the United States' 13.2 percent, as shown in Figure 6-5.

Fig. 6-3. Changes in Japan's Budget for Bilateral Scientific Assistance

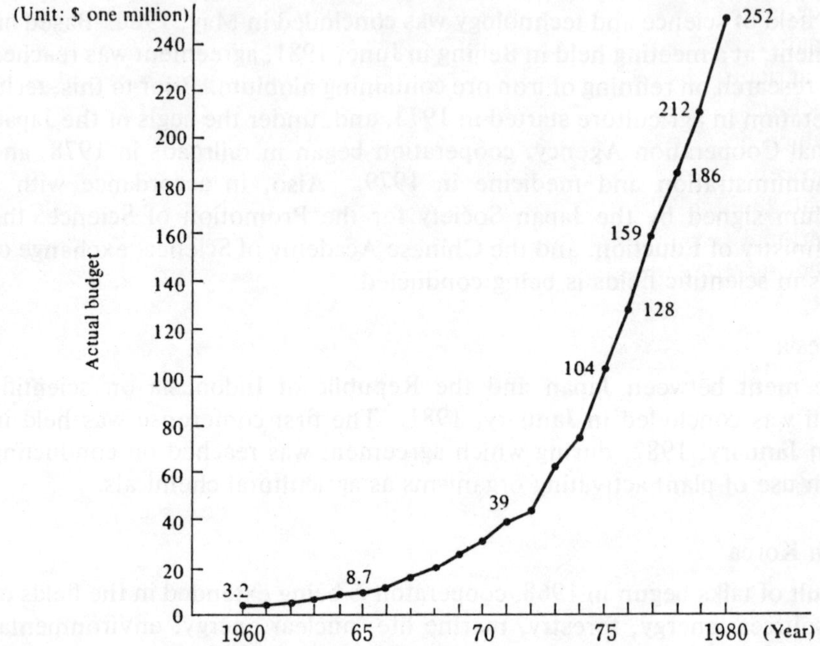

Source: The Ministry of Foreign Affairs

Fig. 6-4. Ratio of Bilateral Assistance Budgets to Total Government Technology Support Funds Among DAC Countries

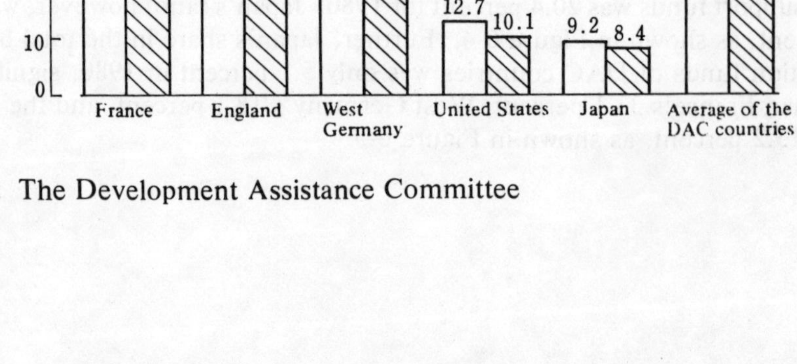

Source: The Development Assistance Committee

Technical assistance in Japan is being carried out both at the governmental and the private levels. At the governmental level, the Japan International Cooperation Agency, subsidized by the Ministry of Foreign Affairs, and the Ministry of International Trade and Industry are providing assistance in a variety of industrial, agricultural, and other projects. The nucleus of governmental assistance in technical training and guidance is the Japan International Cooperation Agency, which accepts trainees, despatches experts, sets up overseas technical cooperation centers, and supplies equipment. On the private level, apart from purely private technological help, assistance is being given with the cooperation of the International Development Center of Japan, the Japan Consulting Institute, the Engineering Consulting Firms Association, the International Engineering Consultants Association, the Japan Transport Consultants Association, and the Japan Telecommunications Engineering and Consulting Services. Various groups conducting activities are shown in Table 6-6.

Fig. 6-5. Share of Technical Assistance of DAC Countries (1980)

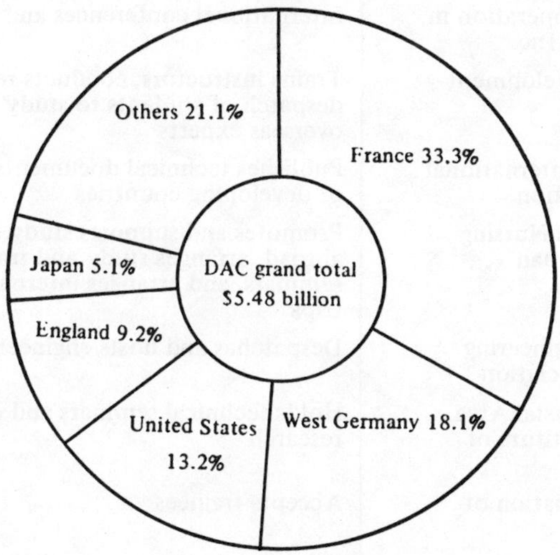

Source: The Development Assistance Committee

Table 6-6. Main Private Groups Offering Technical Assistance

Group	Activity
Asia Kaikan	Constructs, manages, and provides accomodation and welfare facilities for trainees from Asian countries
OISCA Industrial Development Body	Despatches technicians, accepts technical trainees, trains technicians, and holds international conferences and seminars
Association for Overseas Technical Scholarship	Accepts trainees
Overseas Fishery Cooperation Foundation	Supplies funds for international assistance and trains instructors
Japan Telecommunications Engineering and Consulting Service	Hosts technicians, accepts trainees, and organizes training
Overseas Electrical Industry Survey Institute, Inc.	Provides technical assistances
Overseas Agricultural Development Association	Provides consultations, exchanges information, and conducts research
Japanese organization for International Cooperation in Family Planning, Inc	Despatches experts, accepts trainees and holds international conferences and research meetings
International Development Center of Japan	Trains instructors, conducts research, arranges despatch of students to study abroad, and hosts overseas experts
Association for International Technical Promotion	Publishes technical documents in local languages of developing countries
The International Nursing Foundation of Japan	Promotes and supports study of students abroad, arranges study and training, holds seminars, and arranges international exchange trips
International Engineering Consultants Association	Despatches and hosts engineers and instructors
The Overseas Coastal Area Development Institute of Japan	Holds technical seminars and conducts research
The I.L.O. Association of Japan	Accepts trainees
International Medical Foundation of Japan	Operates the Southeast Asia Medical Information Center, accepts trainees, and assists studies abroad
Japan Overseas Development Corporation	Despatches experts in production engineering and management

Source: The Ministry of Foreign Affairs

Next, Figure 6-7 looks at the ratio of the Japanese government's technological assistance by area. In 1980 the percentage for Asia was overwhelmingly large at 49.7 percent of the whole, followed by 19.9 percent for Latin America, 11.5 percent for Africa, and 8.7 percent for the Middle East.

Fig. 6-7. Ratio of Japanese Government's Technological Assistance by Area (1980)

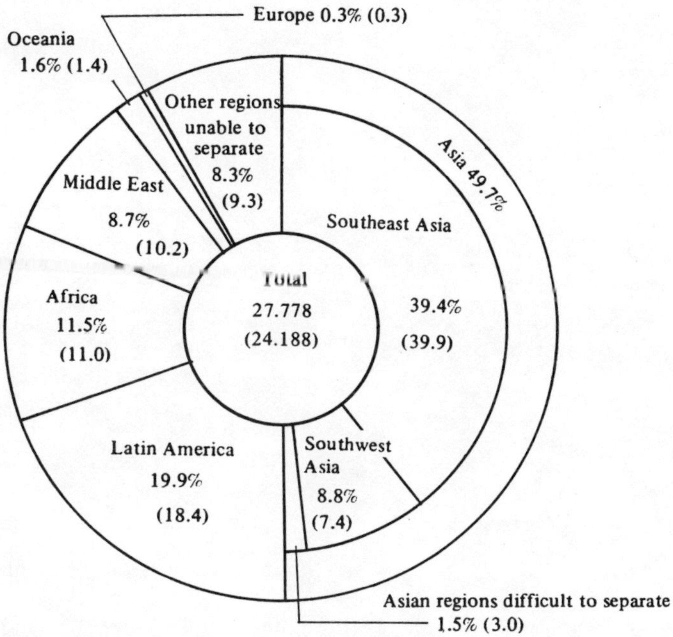

Note: Figures in parentheses indicate percentage for the previous year
Source: The Ministry of Foreign Affairs

Although developing countries need technology in order to grow, lack of sufficient funds and technical personnel present major obstacles. As pointed out by the United Nations, industrialized countries should cooperate closely with developing countries in providing or developing technology suited to each particular country. Through various projects, the Japanese government is making efforts to contribute its fair share toward achieving this goal.

Next Figure 6-7 looks at the ratio of the Japanese government's technological assistance by area. In 1980 the percentage for Asia was overwhelmingly large at 49.7 percent of the whole, followed by 19.9 percent for Latin America, 11.0 percent for Africa, and 8.7 percent for the Middle East.

Fig. 6-7 Ratio of Japanese Government's Technological Assistance by Area (1980)

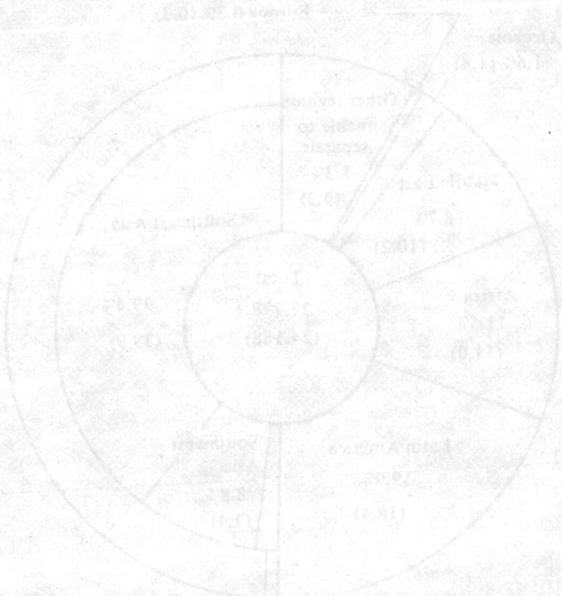

Note: Figures in parentheses indicate percentage for the previous year.
Source: The Ministry of Foreign Affairs

Although developing countries need technology in order to grow, lack of sufficient funds and technical personnel present major obstacles. As pointed out by the United Nations, industrialized countries should cooperate closely with developing countries in providing or developing technology suited to each particular country. Through various projects, the Japanese government is making efforts to contribute its fair share toward achieving this goal.

CHAPTER 7

ALLOCATIONS FOR GOVERNMENT RESEARCH

As estimated by the Science and Technology Agency, Japan's science and technology budget in 1981 was about $5.6 billion, up 8.2% over the previous year. In the same year, however, the growth rate of the national general accounts budget, at 9.9 percent, was slightly higher. However, the growth rate of the general annual expenditure in that year (the general account minus both government bond expenditures and grants to local government) was 4.3 percent, and compared with this figure the 8.2 percent increase in the science and technology budget was about double.

Figure 7-1 shows changes in the science and technology budget for individual items. Of these items, the subsidy account has been on the increase (up 10 percent in 1981 over the previous year) because of growth in the size and number of large-scale projects undertaken by various research organizations, both private and semi-governmental. As a result, the proportion of the subsidy account in the total science and technology expenditure increased from 42.0 percent in 1977 to 49.0 percent in 1981.

Fig. 7-1. Changes in Science and Technology Budget by Item

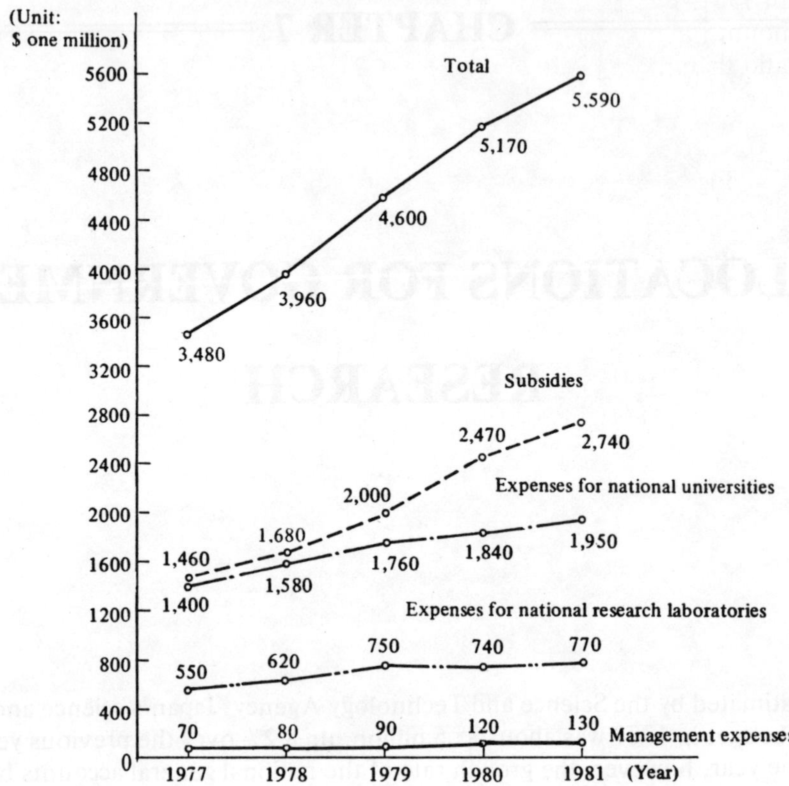

Within the science and technology budget, research expenses in the science promotion budget and the energy-related budget together amounted to about $2.2 billion in 1981, up 5.6 percent over the previous year, as shown in Figure 7-2. This amount represents 11.8 percent of the national general accounts budget, a lower ratio than in the previous year.

Fig. 7-2. Changes in the Science and Technology Budget

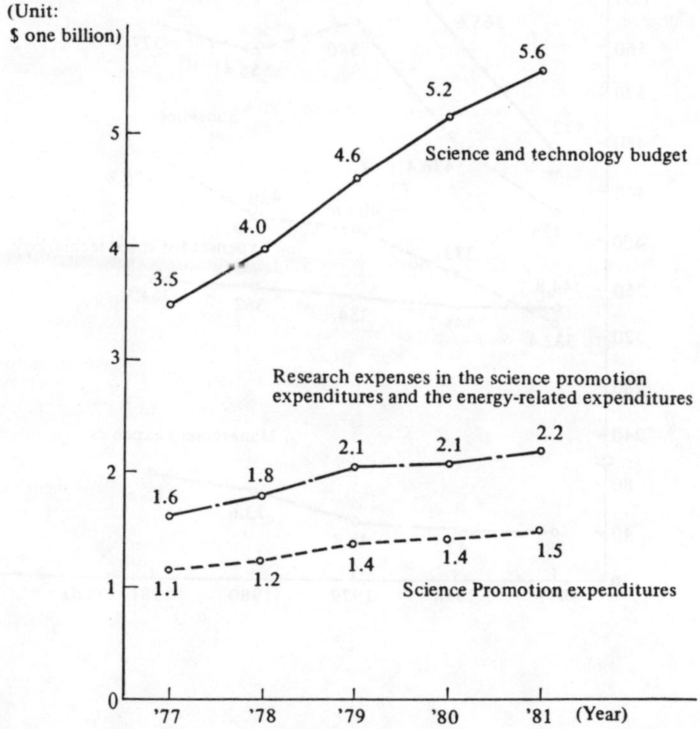

Next, Figure 7-3 shows changes in individual research expenses within the science promotion budget and the energy-related budget. In 1981 the amount for nuclear power was $708 million (up 4.7 percent over the previous year), for space technology, $365 million (up 0.8 percent), and for public research laboratories, $577 million (up 3.7 percent).

Table 7-4 presents a breakdown of the science and technology expenditure as allotted to individual ministries and agencies.

Fig. 7-3. Changes in Research Expenses within the Science Promotion Budget and the Energy-related Budget

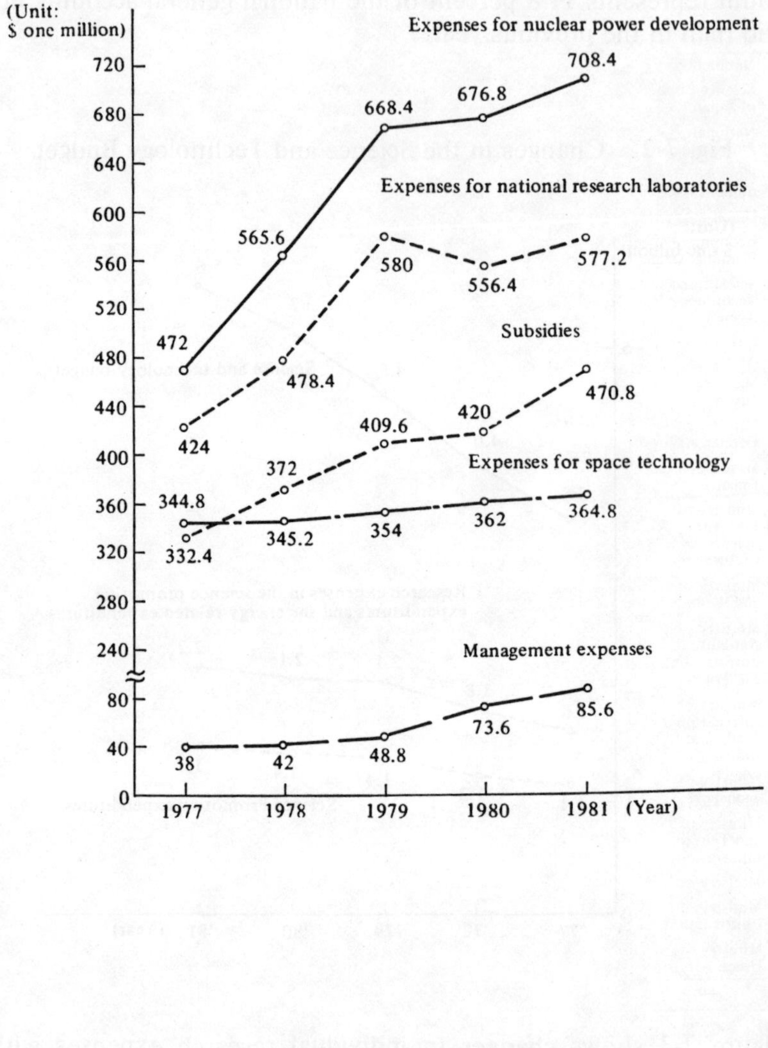

Table 7-4. Science and Technology Budget by Ministry and Agency

(Units: $ one million)

Year / Ministry and Agency	1980				1981			
	A	B	C	B+C	A	B	C	B+C
National Diet	1.9	1.9	–	1.9	2.0	2.0	–	2.0
Science Council of Japan	–	–	3.3	3.3	–	–	2.9	2.9
National Police Agency	2.8	2.8	–	2.8	2.9	2.9	–	2.9
Hokkaido Development Agency	0.6	0.6	4.7	5.3	0.6	0.6	4.0	4.6
Defense Agency	–	–	118.4	118.4	–	–	130.3	130.3
Economic Planning Agency	2.7	2.7	–	2.7	2.9	2.9	–	2.9
Science and Technology Agency	529.8	1161.3	178.3	1339.6	576.0	1234.9	218.6	1453.5
Environment Agency	36.0	36.0	6.7	42.7	37.8	37.8	7.3	45.1
Ministry of Justice	2.8	2.8	–	2.8	2.9	2.9	–	2.9
Ministry of Foreign Affairs	–	7.4	8.1	15.5	–	8.3	8.3	16.6
Ministry of Finance	1.3	1.3	2.0	3.3	1.3	1.3	2.0	3.3
Ministry of Education, Science and Culture	209.5	209.5	2359.5	2569	231.8	231.8	2516.2	2748.0
Ministry of Health and Welfare	77.4	77.4	23.1	100.5	81.7	81.7	27.3	109
Ministry of Agriculture, Forestry and Fisheries	212.0	212.0	12.3	224.3	224.6	224.6	12.2	236.8
Ministry of International Trade and Industry	244.5	286.0	343.7	629.7	252.2	292.2	440.1	732.3
Ministry of Transport	42.4	42.4	6.3	48.7	40.1	40.1	6.9	47.0
Ministry of Posts and Telecommunications	22.2	22.2	–	22.2	18.9	18.9	–	18.9
Ministry of Labor	2.1	2.1	3.1	5.2	2.2	2.2	3.3	5.5
Ministry of Construction	18.4	18.4	10.0	28.4	19.2	19.2	6.9	26.1
Ministry of Home Affairs	1.9	1.9	–	1.9	2.0	2.0	–	2.0
Total	1408.3	2088.7	3079.5	5168.2	1499.1	2206.3	3386.3	5592.6

Note: A: Science promotion budget
B: Research expenses in the science promotion budget and the energy related budget
C: Non-research expenses in the science promotion budget and the energy-related budget

CHAPTER 8

GOVERNMENT PROMOTION OF RESEARCH

In Japan, the role played by national laboratories, semi-government laboratories, and universities has become increasingly important in meeting ever-changing social demands for progress in science and technology.

This chapter describes the promotion of research activities in these three sectors, as well as the present status of major interdisciplinary projects.

1. OUTLINE OF ACTIVITIES

(1) National Laboratories

National laboratories, each belonging to a particular ministry or agency, had total expenditures of $768.5 million in 1981, up 4.1 percent over the previous year, broken down by ministry and agency in Table 8-1. The number of personnel working in these organizations to talled 16,096 in 1981, nearly the same as the previous year, and personnel expenses accounted for about 50 percent of total expenditures, while research expenses accounted for about 35 percent.

Research at national laboratories can be divided into two kinds, on-going basic research and limited-term applied research. Expenditures for basic research, which provides a foundation for other research, amounted to about $140 million in 1981, down 2.8 percent from the previous year. Expenditures for applied research, which is conducted according to social and economic needs and must be

Table 8-1. Expenditures for National Research Laboratories

(Unit: $One million)

Ministry and Agency \ Year	1980 A	1980 B	1980 A+B	1981 A	1981 B	1981 A+B	Laboratories
National Police Agency	2.8	–	2.8	2.9	–	2.9	National Research Institute of Police
Hokkaido Development Agency	0.6	4.7	5.3	0.6	4.0	4.6	Civil Engineering Research Institute
Defense Agency	–	115.1	115.1	–	126.8	126.8	Technical Research and Development Institute
Economic Planning Agency	2.7	–	2.7	2.9	–	2.9	The Economic Research Institute
Science and Technology Agency	77.5	–	77.5	91.4	–	91.4	National Aerospace Laboratory and others
Environment Agency	31.2	–	31.2	33.1	–	33.1	National Institute for Environmental Pollution Research and others
Ministry of Justice	2.8	–	2.8	2.9	–	2.9	Research and Training Institute of Justice
Ministry of Finance	1.2	1.7	2.9	1.3	1.8	3.1	Research Institute of Brewing and others
Ministry of Education, Science and Culture	28.7	–	28.7	29.0	–	29.0	National Research Institute of Genetics and others
Ministry of Health and Welfare	36.1	7.7	43.8	36.2	10.5	46.7	National Institute of Health and others
Ministry of Agriculture, Forestry and Fisheries	194.7	0.1	194.8	206.8	–	206.8	Upland Farming Research Center and others
Ministry of International Trade and Industry	143.1	–	143.1	138.9	–	138.9	National Research Laboratory of Metrology and others
Ministry of Transport	29.6	2.7	32.3	26.8	3.4	30.2	Ship Technical Research Institute and others
Ministry of Posts and Telecommunications	22.2	–	22.2	18.8	–	18.8	Radio Research Laboratory and others
Ministry of Labor	2.1	2.2	4.3	2.2	2.4	4.6	Industrial Safety Institute and others
Ministry of Construction	16.3	10.0	26.3	17.1	6.7	23.8	Public Works Research Institute and others
Ministry of Home Affairs	1.9	–	1.9	2.0	–	2.0	Fire Defense Research Institute and others
Total	593.5	144.2	737.7	612.9	155.6	768.5	

Note: A: Research expenses in the science promotion budget and the energy-related budget

B: Research expenses other than the above

Table 8-2. Breakdown of Applied Research Budget by Ministry

(Units: $one thousand)

Ministry and Agency \ Item \ Year	1978 Number of projects	1978 Budget	1979 Number of projects	1979 Budget	1980 Number of projects	1980 Budget	1981 Number of projects	1981 Budget
National Police Agency	3	140	3	140	2	136	2	104
Science and Technology Agency	19	7,220	20	11,552	20	15,432	20	24,872
Environment Agency	6	1,464	8	1,940	10	2,544	11	2,756
Ministry of Justice	1	4	1	4	1	4	1	4
Ministry of Finance	1	116	1	136	1	136	1	148
Ministry of Education	29	548	28	708	27	640	26	544
Ministry of Health and Welfare	9	228	9	232	9	232	9	220
Ministry of Agriculture, Forestry and Fisheries	26	2,816	26	2,684	29	2,844	29	2,688
Ministry of International Trade and Industry	122	11,528	118	12,204	128	12,248	128	11,720
Ministry of Transport	19	4,008	20	2,016	17	1,692	18	1,192
Ministry of Posts and Telecommunications	9	11,640	9	9,940	11	9,428	12	5,652
Ministry of Labor	3	196	4	248	5	328	5	324
Ministry of Construction	5	212	4	180	4	140	3	88
Ministry of Home Affairs	4	180	3	196	3	192	3	192
Pollution control (Environment Agency)	115	12,256	117	12,992	117	12,992	117	12,992
Nuclear power (Science and Technology Agency)	192	4,948	203	5,976	120	6,536	114	6,952
Total	563	58,104	573	61,148	504	65,524	499	70,448

Note: The above figures are those contained in the science promotion budget

accomplished in a definite period of time, amounted to about $73 million in 1981, up 0.9 percent over the previous year. The number of these research projects totalled 499 in 1981, broken down by ministry and agency in Table 8-2.

(2) Semi-government Laboratories

These laboratories, funded by both the government and private industry, together with national laboratories play an important role in public research activities. Because laboratories of this type can fund and gather researchers from both public and private sectors, operate flexibly, and can efficiently conduct research aimed at achieving specific goals, their role has grown increasingly important as research projects become ever larger and more complex.

Table 8-3 shows changes in the amounts of government funds and subsidies given to various semi-government laboratories.

Table 8-3. Government Funds and Subsidies for Semi-government Laboratories

(Units: $One million)

Organizations	Year	1977	1978	1979	1980	1981
Institute of Physical and Chemical Research	Fund	9.1	11.0	13.1	15.4	14.7
	Subsidy	15.5	17.1	17.7	18.4	19.3
Japan Atomic Energy Research Institute	Fund	105.3	131.2	180.0	210.5	251.9
	Subsidy	48.9	53.9	57.3	61.6	65.8
Japan Nuclear Ship Development Agency	Fund	2.4	3.7	10.8	20.4	21.7
	Subsidy	4.6	5.0	5.5	5.4	5.7
Power Reactor and Nuclear Fuel Development Corporation	Fund	232.3	278.7	313.6	416.2	428.0
	Subsidy	46.0	55.1	58.8	68.4	75.7
National Space Development Agency of Japan	Fund	296.2	295.3	303.1	308.9	313.0
	Subsidy	20.0	22.4	24.2	26.0	28.4
Marine Science and Technology Center	Fund	2.0	8.5	12.4	15.0	23.6
	Subsidy	2.3	2.7	2.9	3.2	3.5
Institute of Agricultural Machinery	Fund	0.7	0.6	0.6	0.4	0.3
	Subsidy	2.1	2.2	2.4	2.5	2.6

(3) Universities

Most academic research at universities is directed at exploring the unknown through a steady, long-term accumulation of fundamental scientific knowledge, and at creating original scientific breakthroughs.

Such research is conducted at national, local-governmental, and private universities, and at research institutes attached to these universities. Although university budgets combine appropriations for both research and teaching, making it difficult to isolate research expenditures, government appropriations for academic research can be classified into those for standard, on-going research and special appropriations for specified research projects.

The appropriations for standard, on-going research support the fundamental work of each researcher, and those for national universities include such items as personnel expenses and travel expenses. Subsidies are also given to private universities to support their educational and research activities.

Among the various special appropriations to universities for specified research projects, of special note is the scientific research subsidy, intended to achieve a high degree of scientific progress, and given to specified researchers whose choosen work is deemed to be of special significance to the country's level of academic research. This system has been very successful in creating a large number of original and innovative academic achievements and in producing first-rate research groups in various fields. An outline of these scientific research subsidies and their budgets is given in Table 8-4.

Table 8-4. Scientific Research Subsidies (1981)

(Unit: $One million)

Item	Note	Budget
Special Research	Long-term research in fields of extremely high scientific or social demand, such as cancer treatment, disasters prevention, energy development, and environment control	22.6
Specified Research	Short-term research in fields of extremely high scientific or social demand, such as life and nuclear sciences (normally three years)	19.08
Comprehensive Research	Joint research among different research organizations	10.84
General Research	Research conducted by a researcher, or researchers in an organization	55.56
Experimental Research	Applied research for putting basic research into practical application	10.12
Encouragement of Research (A)	Research conducted by a researcher with relatively short experiences (Expense: less than $4,000)	8.92
Encouragement of Research (B)	Research conducted by elementary, middle, or high school teacher or by a citizen not belonging to any research organization (Expense: less than $1,200)	0.36
Publishing Expense	Expense to support publication of research results with scientific value	3.12
Overseas Research	Survey, mainly field survey conducted in a foreign country	12.6
	Total	143.2

(4) Special Promotion Systems

As described in Chapter 2, there are two special systems for promotion of research aimed at creative, independently-developed technology, the System for Exploratory Research for Advanced Technology, and the Research and Development Project of Basic Technology for Future Industries. As of 1981, four projects have been initiated under the first system, and twelve under the second. These projects and their budgets are shown in Tables 8-5 and 8-6.

Table 8-5. Outline of Research Projects under the System for Exploratory Research for Advanced Technology

Project	Appropriated Budget	Outline
Ultra-fine particles	$640,000	A study of the peculiar properties of ultra-fine particles (less than 0.1 micron), distinctly different from solid materials, and search for their possible applications
Amorphous & laminar materials	$724,000	A study of the structures and properties of amorphous and laminar materials from physical, mechanical, metallurgical, and electronic points of view in order to develop new materials and elements
Fine polymers	$400,000	A study of the synthesis of polymers with selective permeability or semiconductivity
Perfect Crystals	$636,000	A study of perfect crystalization technology and the electrostatic induction effect in order to create new semiconductor elements

Table 8-6. Outline of Research Projects under the Research and Development Project of Basic Technology for Future Industries

Project	Outline
New Materials ($5,424,000) 1. High performance ceramics	o Development of structural or engineering ceramics which are hard, rust-free, dimensionally accurate, and suitable for high-temperature applications. o The ceramics is expected to be applied in such high-tech fields as nuclear power, new energy, and aerospace, as well as in improving the efficiency of energy-saving thermal engines.
2. Synthetic membrane for new separation technology	o Development of synthetic membranes which by utilizing difference in the property of matters freely separate and refine mixed gases or liquids. o The membranes, when developed and applied to separation and refining processes at the chemical industry, will materialize energy-saving and pollution-free factories.

Table 8-6. continued

Project	Outline
3. Synthetic metals	o Development of synthetic metals (polymer materials) which, unlike real metals, are light-weight and corrosion resistant, but still electrically conductive. o The metals will replace copper and aluminum used in electrical wires and cables, and also are good candidate as super-conductive material.
4. High performance plastics	o Development of high performance plastics, light but metal-strong polymers, using high crystalization technique. o The plastic are expected to replace aluminum and steel as structural materials, and are also considered as super-strength materials which are elastic and electrically insulating.
5. Advanced alloys with controlled crystaline structures	o Development of strong alloys, exceeding the performance of conventional alloys, through single crystalization and grain refining techniques. o The alloys will improve the reliability of nuclear reactors, aircrafts, and aerospace equipments, and also contribute to energy conservation.
6. Advanced composite materials	o Development of advanced composite materials, lighter than aluminum alloys and stronger than steel, with specified strength and hardness for highly reliable structural materials. o The composite materials are considered indespensable for further development of machineries such as aircrafts, cars, space equipments, and their lightness will also contribute to energy conservation.
[Bio-technology] ($2,700,000)	
1. Bio-reactor	o Development of bioreactor technology for new chemical reactions that uses micro-organisms or enzymes. o The technology which offers reaction under normal temperature and pressure is advantageous over conventional chemical reactions, and can realize energy-saving and pollution-free chemical industry.
2. Large scale cell cultivation	o Development of a method to replace a serum, such as from unborn calf, presently supplemented in medium to support proliferation of mammalian cells in culture, for various cell lines and make submerged cultures possible. This is one of the essential basic technologies for the bio-industry. o The method for a serum-free medium and/or submerged culture will drastically reduce the cost of large-scale cultivation and time for purification of products.
3. Utilization of recombinant DNA	o Development of DNA recombinant technology, another basic technology essential to bio-industry, could be used not only to obtain hormones and enzymes, but also to produce bulk commodity chemicals. o The technology makes it possible to create new micro-organisms to be used in the production process of chemicals.
[New electronic devices] ($2,692,000)	
1. Super-lattice devices	o Development of super-lattice devices that have an extremely fine structure tailored to atomic scale, and utilize new electronic effects arising from such a structure. o The device is expected to realize ultra-high frequency and ultra-high speed operation at room temperature.
2. Three-dimensional IC's	o Development of three dimentional IC's by alternately stacking semiconductor layers for active elements and insulator layers for separation. o The IC's will be ultra-large-scale and monolithic integration of various functional elements such as for sensors, processors, memories, and displays to realize further miniaturization of computers.
3. Fortified IC's for extreme conditions	o Development of fortified IC's that can withstand severe conditions such as high energy irradiations, high temperatures, mechanical vibrations, and shocks, to construct electronic systems reliably functioning under extreme conditions. o The IC's are expected to expand the usage of computers in various new areas including space development and nuclear power generation.

2. INTERDISCIPLINARY PROJECTS

Research projects in recent years have become more intricate and diversified, calling for systematic, interdisciplinary cooperation among various research fields. Such large projects as the development of nuclear and non-nuclear energy and of space and marine technology all require enormous budgets, time, and manpower, making close cooperation between public, private, and academic sectors indispensable. In addition, vital new scientific fields such as life sciences, disaster-prevention, and the software sciences are also greatly dependent upon interdisciplinary cooperation.

(1) Nuclear Energy

Under a long-term plan approved in June, 1982, by the Atomic Energy Commission on utilization of nuclear energy, Japan has been stepping up research on nuclear energy as an alternative to fossil fuel energy.

For progress in the utilization of nuclear energy, it is first necessary to establish high standards of safety and to gain the understanding and cooperation of the public. For this purpose, research on a variety of safety measures is being conducted under the guidance of the Nuclear Safety Commission, in addition to the administration of relevant codes by the appropriate ministries.

Currently, there are 23 nuclear power plants in Japan, producing 16 million kW of electricity or nearly 12 percent of the country's total electric power, and thus representing a vital source of power supply. In April, 1982, the government announced a plan to increase nuclear power production capacity to some 46 million kW by April, 1990.

To achieve this goal and also to utilize nuclear energy more effectively, research on safety measures, an independent nuclear fuel cycle, new reactors, and a variety of other subjects is being conducted.

Total nuclear energy related expenditures in Japan were about $1 billion in 1980 and $1.1 billon in 1981. The 1980 figure, about one-third of that of the United States, was at approximately the same level as France and West Germany.

(a) Safety Measures

In order to utilize nuclear energy safely, research on safety measures has been given top priority in technological development from the very earliest stage of work, and unprecedentedly stringent legal measures regarding safety have been maintained, covering installation, operation, and inspection.

The Japan Atomic Energy Research Institute is conducting research on the safety of light-water reactors. To study accidents such as those related to the degree of nuclear reaction and to loss of coolant, tests using the NSRR (Nuclear Safety Research Reactor) and experiments on the effectiveness of emergency core cooling systems using the ROSA (Rig of Safety Assessment) were carried out in 1981.

To reduce discharge of radioactive material from nuclear energy facilities to the environment, development of recovery and processing techniques for the used coolant are also being studied, and efforts are being made to lower the level of discharge well below the limits set by the International Commission on Radiological Protection (ICRP). At the National Institute of Radiological Sciences, the effects of exposure to radiation on the human body and heredity are being studied.

Measures to prevent nuclear accidents are being studied and planned based on disaster prevention regulations. Following the accident at the Three Mile Island Nuclear Power Plant in the U.S. in 1979, based on a 1982 report on safety measures by the Japanese Nuclear Safety Commission the government issued various manuals to provide advice and guidance to local governments.

(b) Establishment of a Nuclear Fuel Cycle

With the increasing use of nuclear energy for generation of electric power, the establishment of an independent nuclear fuel cycle, stable supplies of uranium, and effective use of atomic energy are becoming increasingly important.

Japan, poor in uranium resources, is dependent on foreign sources, and to secure a stable supply must make long-term purchase contracts with diversified supply sources. Also, to increase the rate of natural uranium supply, participation in prospecting and mining development projects has become active. In 1981, for example, the Power Reactor and Nuclear Fuel Development Corporation, independently or with joint foreign collaboration, carried out investigations and prospecting expeditions in Australia, Canada, and Africa. In the private sector, overseas uranium prospecting and mining is being promoted with financing from the Metal Mining Agency of Japan.

Stable supplies of enriched uranium are also very important, since the mainstay of the present atomic energy electric power supply, the light-water power reactor, uses enriched uranium as a fuel. For this purpose, Japanese electric power supply companies have contracted with the United States Department of Energy for an assured supply of enriched uranium sufficient to produce a 51 million kW of power annually for 30 years. Also, to produce enriched uranium domestically, centrifuge process technology is being vigorously developed as a national project. In 1980, the Power Reactor and Nuclear Fuel Development Corporation completed construction of a pilot plant with 7000 centrifuge processors, which started full-scale operations in March, 1982. Reprocessing of used fuel is presently being handled by the Tokai Reprocessing Plant, as well as at British and French reprocessing plants. Under these arrangements, an adequate supply is assured up to about 1990.

The Tokai Reprocessing Plant, first test-operated in July, 1977, started full-scale operations in January, 1981. Nevertheless, to cope with increasing demands for reprocessing in the future, a larger reprocessing plant is needed, and in March, 1980, a private company, the Japan Nuclear Fuel Service Co., Ltd., was established for construction and administration of additional reprocessing facilities.

As the last stage of the nuclear fuel cycle, measures for disposal of radioactive waste are also an important matter. Serious efforts are being made to further reduce the level of wastes from atomic energy electric power plants. At the same time, offshore and onshore disposal is being considered, for which both test disposal to assure sufficient safety and public understanding are necessary. For highly-radioactive wastes from reprocessing plants, disposal technology is being researched mainly by the Power Reactor and Nuclear Fuel Development Corporation and the Japan Atomic Energy Research Institute.

(c) New Reactors

Because fast breeding reactors can produce new fuel from spent fuel, they greatly lower the amount of natural uranium necessary in the nuclear fuel cycle. As this increases the effective utilization of uranium resources, various countries around the world are now competitively developing the necessary technology. In Japan, independent research on fast breeder reactors has been carried out since 1968 by the Power Reactor and Nuclear Fuel Development Corporation, with completion expected by the year 2010. Additionally, research is being conducted on development of advanced thermal converter reactors, capable of more efficient use of nuclear fuel. With practical utilization estimated to be achieved around 1990, they are expected to fill the gap between now and the more distant era of the fast breeding reactor.

Operation of the experimental fast breeding reactor "JOYO" with 75,000 kW of heat output has been accomplished, and preparation for 100,000 kW heat output run has started. A prototype reactor, "MONJYU", with about 280,000 kW of electrical output, has passed safety tests by the Administrative Agency and, while construction is awaiting final approval by the local community, relevant research on sodium for the coolant, steam generation, fuel, and safety is now progressing. A prototype of the advanced thermal converter reactor "FUGEN", with 165,000 kW of electrical output, started operations in March, 1979. In November, 1980, however, the cooling system failed due to corrosion, requiring one year to restart operations.

At that time, a portion of the fuel was replaced for the first time with plutonium processed at the Tokai Reprocessing Plant, thus completing an entirely domestic nuclear fuel cycle. An evaluation committee set up in 1980 by the Atomic Energy Commission issued a report on the technological and economic aspects of constructing a reactor with an output of 600,000 kW of electricity, which in 1981 became the basis for discussions between the Atomic Energy Commission and the Federation of Electric Power Companies.

(d) Multi-purpose High-Temperature Gas Reactor

In light of the energy shortage, high expectations have been put on the use of nuclear heat energy as a substitute for oil. With increased interest in the multi-purpose gas reactor as a source of industrial energy, the Atomic Energy Commission has been actively promoting research, and plans have been made for an experimental reactor to start operations by 1990. The Japan Atomic Research Institute, having conducted research on a multi-purpose high-temperature gas reactor since 1969, in 1981 successfully test-operated the main section of the helium engineering demonstration loop (HENDEL) at about 1000°C. Construction of the remaining equipment is still going on. Detailed designing of the experimental reactor was started in 1969 and is advancing along with research on reactor physics, controls, and fuels.

(e) Nuclear Fusion

Research and development of nuclear fusion is progressing on the basis of the Atomic Energy Commission's second fundamental plan set in July, 1975, which places emphasis on achievement of the critical plasma conditions necessary for actualization of the fusion reactor. Research is focused on achievement of critical plasma conditions with JT-60 equipment, development of a non-circular cross-section plasma Tokamak device, high beta plasma, plasma diagnosis, and fusion reactor technology. This research is being carried out by the Japan Atomic Energy Research Institute, the Electrical Laboratory of MITI, the National Research Institute for Metals, and the Institute of Physical and Chemical Research.

Other than the above institutes, research on nuclear combustion, alternate systems to Tokamak, and reactor materials is being carried out at universities, such as Nagoya University, Kyoto University, Osaka University, and Tsukuba University.

(f) Radiological Science

Radiation is used in a wide variety of fields, from fundamental sciences to practical applications like medicine, agriculture, and industry. The number of enterprises using radio isotopes and various radiation-generating equipment is increasing year by year, and as of March, 1982, the total number of enterprises which have received or have applied for use of radio-active materials totalled 4,144.

In the medical field, radiation is used for X-ray diagnosis, treatment with high radiation, and diagnostical treatment with radio isotopes. For treatment of malignant tumors with radiation therapy, use of cyclotron equipment is drawing attention and has been added to existing therapeutic methods. The medical cyclotron used at the National Institute of Radiological Science since February, 1976, is showing remarkable therapeutical effects on malignant black tumors.

In the industrial field, the diversified uses of radiation include such fields as measurement, testing, and analysis. In the agricultural field, radiation is used to improve crops and as a tracer for the study of physiological structures of marine, agricultural, and forestry products.

Together with progress in the use of radiation, consumption of isotopes is increasing year by year. For the supply of these isotopes, besides imports, research laboratories such as the Japan Atomic Energy Research Institute are working on domestic production. Production of short-life nuclear materials, difficult to import, is being studied at the National Institute of Radiological Science and the Institute of Physical and Chemical Research.

(2) Non-nuclear Energy

Non-nuclear energy sources, such as coal, solar, and geothermal energy, are also necessary to secure a stable supply of energy. Currently, research is being actively carried out in Japan on fossil fuels, natural energy, and more effective use of energy.

A number of new and advanced technologies, such as gasification and liquefaction of coal, have been developed to produce cleaner exploitation of this form of energy. A small-scale test plant for production of liquid and gas coal has been completed and is ready for test operations.

In the natural energy area, research in geothermal, solar, ocean, wind, and biomass energy is being conducted in hopes of increasing the domestic supply of energy.

Geothermal power generation, using hot water or vapor in volcanic regions, is an ideal power source for Japan, a volcanic country. Electrical output geothermal energy amounts to about 160,000 kW in Japan, and to further increase this, exploration of sites and development of improved technology is being conducted.

Research on use of solar energy for heating, cooling, and hot water supply systems is also advancing. At the same time, as an electric power source, solar heat and photo-electric generation systems are also being researched, and currently two types of pilot power plants using solar heat, each with a capacity of 1,000 kW, have been constructed and are undergoing tests.

Research is also being carried out on utilization of wave energy, temperature differences in the ocean, and wind power. Wave energy, for example, has been converted to mechanical energy by utilizing the force of moving waves to compress air in an enclosed space, much like the functions of a piston.

Energy can also be converted from the photosynthesis mechanism of living organisms in a process called bio-conversion. Currently, research is advancing on plants producing hydrocarbons, the photosynthesis mechanism for production of hydrogen, and organic solar cells.

To utilize existing energy more effectively, work is progressing in the development of various new technologies to produce and store electrical power, the use of hydrogen as an energy source, and recycling of waste for the generation of energy.

(3) Space Technology

With recent technological advances, space has become a new frontier for man, vital to future progress. Satellites are increasingly being used for exploration, observation, and communications.

Space development in Japan is carried out under the administration of the National Space Development Agency of Japan and the Institute of Space and Astronautical Sciences. The former is in charge of development of satellites for applications such as communications and meteorology, and the latter for development of satellites for scientific research. Since launching the first satellite, "Ohsumi," in 1970, as of March, 1982, together they had launched 23 satellites.

In the field of scientific research, various observations, such as of X-ray stars and solar flares, are being made by the scientific satellites No. 6, "Jikiken," launched in September, 1978, No. 4, "Hakucho," in February, 1979, and No. 7, "Hinotori," in February, 1981. Currently, scientific satellites No. 8 (ASTRO-B), No. 9 (EXOS-C), No. 10 (PLANET-A), No. 11 (ASTRO-C), and a satellite for SEPAC (Space Experiments with Particle Accelerators) are under development.

The meteorological satellite "Himawari" and the ionosphere survey satellite "Ume 2" were launched in 1977 and have been used for observations. In 1981, a second meteorological satellite, "Himawari 2," was launched with an N-II rocket. At present, marine observation satellite No. 1 (MOS-1) is under development.

Experiments using the communications satellite, "Sakura," launched in December, 1977, and the broadcasting satellite, "Yuri," launched in April, 1978, have been made. In addition, communications satellites No. 2 (CS-2a and CS-2b) and broadcasting satellites No. 2 (BS-2a and BS-2b) are being actively developed.

Development of satellite-launching rockets is also being actively pursued. The institute of Space and Astronautical Sciences developed the M 3S rocket, equipped with thrust vector control units on the first stage, as a vehicle for scientific satellites No. 8 (ASTRO-B) and No. 9 (EXOS-C). To launch scientific satellites No. 10 (PLANET-A) and No. 11 (ASTRO-C), the institute also developed the M-3S II rocket, a modified version of the M-3S.

The National Space Development Agency of Japan, on the other hand, developed the N-1 rocket for experimental technology satellite III (ETS-III), and the N-II rocket, with boosting capacity of 350Kg, for meteorological satellite No. 2 (GMS-2), communications satellite No. 2 (CS-2a and CS-2b) and broadcasting satellite No. 2 (BS-2a and BS-2b). In addition, the agency is developing the H-I rocket, which uses liquid oxygen and hydrogen as a fuel for the second stage and is capable of launching a 550Kg satellite to a geostationary orbit. Tables 8-7 and 8-8 give an outline of developments in satellites and rockets in Japan, and Table 8-9 shows changes in the space technology budget.

Table 8-7. Development of Satellites in Japan

(1) Satellites launched after 1975

Category	Name	Weight (kg)	Altitude of orbit (km)	Launch vehicle	Launch date	Objectives
Scientific research satellites	Experimental satellite,	129	Elliptical orbit 790/3,810	M–3H	2.19.1977	Testing of launch vehicle and satellite attitude control
	Scientific satellite, Tansei 3	126	Elliptical orbit 630/3,970	M–3H	2.4.1978	To observe plasma density, its temperatures, electron energy, and auroral particles
	Scientific satellite No. 6, Jikiken	90	Long elliptical orbit 220/30,100	M–3H	9.16.1978	To observe electron density, corpuscular radiation, and plasma waves
	Scientific satellite No. 4, Hakucho	96	Elliptical orbit 550/580	M–3C	2.21.1979	To observe X-ray star and nebulae
	Experimental satellite, Tansei 4	185	Elliptical orbit 520/610	M–3S	2.17.1980	To test launch vehicle and satellite equipments
	Scientific satellite No. 7, Hinotori	188	Elliptical orbit 580/640	M–3S	2.21.1981	To observe solar X-ray flare, X-ray burst, and corpuscular radiation
Application satellites	Experimental technology satellite I, Kiku	82	Circular orbit 980/1,100	N – I	9.9.1975	To test launch vehicle, satellite tracking control, and antenna
	Ionosphere survey satellite, Ume	139	Circular orbit 990/1,010	N – I	2.29.1976	To observe ionosphere critical frequency and source of radio noise
	Experimental technology satellite II, Kiku 2	130	Geostationary orbit 35,800	N – I	2.23.1977	To test launch and tracking control
	Geostationary meteorological satellite, Himawari	315	Geostationary orbit 35,800	Delta rocket of the U.S.	7.14.1977	To make meteorological observation for the GARP (Global Atmospheric Research Program)
	Experimental geostationary communications satellite, Sakura	340	Geostationary orbit 35,800	Delta rocket of the U.S.	12.15.1977	To test operating technology for satellite communication system
	Ionosphere survey satellite, Ume 2	141	Circular orbit 980/1,220	N – I	2.16.1978	To observe ionosphere critical frequency for effective operation of short wave communications
	Experimental broadcasting satellite, Yuri	355	Geostationary orbit 35,800	Delta rocket of the U.S.	4.8.1978	To test operation of satellite broadcasting system
	Experimental geostationary communications satellite, Ayame	130	Failed	N – I	2.6.1979	To test launch and tracking control of geostationary satellite, and also millimeter wave communication

Table 8-7. continued

	Name	Weight (kg)	Altitude of orbit (km)	Launch vehicle	Launch date	Objectives
Application satellites	Experimental geostationary communications satellite, Ayame 2	130	Failed	N – I	2.22.1980	To test launch and tracking control of geostationary satellite, and also millimeter wave communication
	Experimental technology satellite IV, Kiku 3	638	Long elliptical orbit 220/35,820	N – II	2.11.1981	To testing N-II rocket and satellite equipments
	Geostationary meteorological satellite No. 2, Himawari 2	296	Geostationary orbit 35,800	N – II	8.11.1981	To improve meteorological observations

(2) Satellites under development

	Name	Weight (kg)	Altitude of orbit (km)	Launch vehicle	Launch date	Objectives
Scientific research satellites	Satellite for SEPAC (Space experiments with particle accelerators)	–	–	The space shuttle	1983	To observe auroral luminous mechanism, charged particle movement in plasma, and electromagnetic wave excitation
	Scientific satellite No. 8, ASTRO-B	180	Elliptical orbit 350/600	M–3S	1982	To observe X-ray star, galaxy, burst, and nebulae
	Scientific satellite No. 9, EXOS-C	200	Elliptical orbit 300/1,000	M–3S	1983	To observe stratospheric and mesospheric atmosphere, and ionospheric plasma
	Scientific satellite No. 10, PLANET-A	125	Orbit about sun	M–3S II	1985	To observe interplanetary plasma and Halley's Comet
	Scientific satellite No. 11, ASTRO-C	400	Circular orbit 500	M–3S II	1986	To observe various X-ray sources
Application satellites	Experimental technology satellite III, ETS-III	385	Circular orbit 1,000	N – I	1982	To test high-powered satellites and equipment
	Communications satellite No. 2, CS-2a, CS-2b	350	Geostationary orbit 35,800	N – II	1982 (CS-2a) 1983 (CS-2b)	To develop efficient communication system
	Broadcasting satellite No. 2, BS-2a, BS-2b	350	Geostationary orbit 35,800	N – II	1983 (BS-2a) 1985 (BS-2b)	To develop efficient broadcasting system

Table 8-7. continued

	Name	Weight (kg)	Altitude of orbit (km)	Launch vehicle	Launch date	Objectives
Application satellites	Geostationary meteorological satellite No. 3, GMS-3	300	Geostationary orbit 35,800	N – II	1984	To improve meteorological observations
	Marine observation satellite No. 1, MOS-1	750	Synchronous orbit with Sun	N – II	1986	To observe marine phenomena

Table 8-8. Development of Rockets in Japan

(1) Rockets launched after fiscal 1975

Model	Number of stages	Length (m)	Maximum diameter (m)	Weight (ton)	Propellant	Launched satellites
M-3C rocket	3	20.2	1.41	41.6	All stages solid-fuel	Scientific satellite No. 4, Hakucho
M-3H rocket	3	23.8	1.41	48.7	All stages solid-fuel	Experimental satellite, Tansei 3 Scientific satellite No. 5, Kyokko Scientific satellite No. 6, Jikiken
M-3S rocket	3	23.8	1.41	49.5	All stages solid-fuel	Experimental satellite, Tansei 4 Scientific satellite No. 7, Hinotori
N-I rocket	3	32.6	2.44	90.4	1st & 2nd stage liquid-fuel 3rd stage SOB solid-fuel	Experimental technology satellite I, Kiku Iososphere survey satellite, Ume Experimental technology satellite II, Kiku 2 Ionosphere survey satellite, Ume Experimental geostationary communications satellite, Ayame Experimental geostationary communications satellite, Ayame 2
N-II Rocket	3	35.4	2.44	134.7	1st & 2nd stage liquid-fuel 3rd stage SOB solid-fuel	Experimental technology satellite IV, Kiku 3 Geostationary meteorological satellite No. 2, Himawari 2

Table 8-8. continued

(2) Rockets under development

Model	Number of stages	Length (m)	Maximum diameter (m)	Weight (ton)	Propellant	Schedule	Satellites scheduled for launch after 1982
M-3S rocket	3	23.8	1.41	49.5	All stages solid-fuel	1982 1983	Scientific satellite No. 8 (ASTRO-B) Scientific satellite No. 9 (EXOS-C)
M-3SII rocket	3	28.2	1.41	6.1	All stages solid-fuel	1985 1986	Scientific satellite No. 10 (PLANET-A) Scientific satellite No. 11 (ASTRO-C)
N-I rocket	3	32.6	2.44	90.4	1st & 2nd stage liquid-fuel 3rd stage SOB solid-fuel	1982	Experimental technology satellite III (ETS-III)
N-II rocket	3	35.4	2.44	134.7	1st & 2nd stage liquid-fuel 3rd stage SOB solid-fuel	1982, 1983 1983, 1985 1984 1986	Communications satellite No. 2 (CS-2a, CS-2b) Broadcasting satellite No. 2 (BS-2a, BS-2b) Geostationary meteorological satellite No. 3 (GMS-3) Marine observation satellite No. 1 (MOS-1)
H-I rocket	3	40.3	2.44	139.2	1st stage liquid-fuel 2nd stage liquid oxygen and hydrogen 3rd stage SOB solid-fuel	Late 80's	Capacity to launch a 550 kg satellite into a geostationary orbit

Table 8-9. Changes in the Space Technology Budget in Japan

(Unit: $one thousand)

Organization	1980	1981
Science and Technology Agency	340,388	346,672
National Space Development Agency of Japan	334,908	341,424
National Aerospace Laboratory	3,332	3,144
Indemnity for fishery at the launch site	1,664	1,628
Space Activities Commission	180	184
Others	304	288
National Police Agency	432	0
Environment Agency	0	0
Ministry of Agriculture, Forestry and Fisheries		
Fisheries Agency	184	184
Ministry of International Trade and Industry	572	4,412
Ministry itself	0	4,064
Agency of Industrial Science and Technology	572	348
Ministry of Transport	13,892	15,168
Electronic Navigation Research Institute	28	32
Meteorological Research Institute	0	0
Maritime Safety Agency	824	1,604
Meteorological Agency	13,040	13,532
Ministry of Posts and Telecommunications	11,056	8,552
Ministry itself	2,020	3,368
Radio Research laboratory	9,036	5,184
Ministry of Construction	20	20
Ministry itself	12	12
Geographical Survey Institute	8	8
Ministry of Home Affairs	8	20
Ministry of Education	41,472	44,976
Total	408,028	420,004

(4) Marine Technology

The ocean contains a great variety of biological and mineral resources and also occupies a vast area, making its development and utilization an important subject as the world moves toward a new era of marine order. Ocean development requires not only research on oceanography, meteorology, submarine resources, and marine biology, but also development of basic technology for such things as observation and diving.

In Japan, although various matters concerning marine resources are handled by the relevant ministries and agencies, overall policy on ocean development is the responsibility of the Council for Ocean Development, an advisory organ of the Prime Minister. The Council has pointed out the need to develop a manned submarine capable of diving to a depth of 6,000m for deep-sea surveys. As a step toward this target, the submarine "Shinkai 2000," capable of diving to 2,000m, and its mother ship "Natsushima" were completed in 1981 and used for underwater training. Also, in preparation for diving to a depth of 300m without the use of a submarine, various tests have been conducted, including a simulation experiment in saturation diving to a depth of 30m to study high-pressure physiology.

Since Japan is surrounded by sea, the ocean exerts a great influence on the country's economic and social conditions, and research is being conducted in various fields to conserve and improve the ocean environment. To understand the exchange of seawater between the partially land-sheltered Inland Sea and the main body of the Pacific Ocean, both on-the-spot surveys and experiments with hydraulic models are being conducted. Research on red tide prediction was conducted in 1981 in order to cope with an occurrence of red tide and to develop techniques for the recovery of fishing grounds.

Because marine life plays such an important role in food supply in Japan, it is necessary to promote technology for resource cultivation, the creation of fishing grounds, and the exploitation of unused resources. In the field of resource cultivation technology, government aid has been provided to develop mass production of certain marine plants. Likewise, the National Research Institute of Fisheries Engineering and other research laboratories are conducting research on conservation and improvement of fishing grounds, as well as studies of unused deep-sea resources.

The ocean is rich not only in marine life but also in natural resources such as offshore oil, natural gas, and manganese nodules. The utilization of various substances dissolved in seawater is also expected to become possible, and seawater itself is now considered a feasible water resource. Japan, a country with few under ground resources, has made various efforts in this direction. Prospecting for oil and natural gas in Japanese waters and exploratory drilling have been undertaken. An underwater oil production system is being built under a 7-year (1978 - 1984) project to produce oil from an underwater oil field at a depth of 300m. Experiments at a large-sized inverse osmosis facility have been made to convert seawater to fresh water. Construction of a pilot plant designed to recover uranium from seawater has been started. Existing conditions of deep-sea manganese nodules have been investigated in the sea and basic research on a system for mining them has been made.

The ocean also contains energy in various forms such as wave, heat, and tidal energy. The use of these kinds of marine energy as an energy source is considered very important for the future energy supply in Japan. As part of the development of wave energy, research on a wave-activated power generation system has been conducted.

Development of marine technology has become increasingly active in Japan, as shown by the 32 percent annual increase in the government budget for such technology in 1981. Table 8-10 offers a breakdown of the marine technology budgets for 1980 and 1981.

Table 8-10. Japan's Marine Technology Budget

(Unit: $one thousand)

Ministry or Agency	1980	1981	Growth rate of 1981 or 1980
Hokkaido Development Agency	240	240	0%
Defence Agency	6,608	14,180	115%
Science and Technology Agency	39,104	72,884	86%
Environment Agency	4,332	4,204	Δ3%
Okinawa Development Agency	36	20	Δ44%
National Land Agency	1,552	1,536	Δ1%
Ministry of Education	876	1,440	64%
Ministry of Agriculture, Forestry and Fisheries	44,676	40,516	Δ9%
Ministry of International Trade and Industry	43,072	48,848	13%
Ministry of Transportation	34,980	49,244	41%
Ministry of Posts and Telecommunications	628	248	Δ61%
Ministry of Labor	12	1·2	0%
Ministry of Construction	5,500	6,280	14%
Total	181,616	239,652	32%

(5) Life Sciences

Life sciences, combining studies of biology, medicine, agriculture, chemistry, physics, and engineering, are expected in the near future to bring about epoch-making technological innovations in various fields such as medical treatment, agriculture, industry, and environmental protection.

In Japan, the government has actively promoted policies in support of life sciences since the Council for Science and Technology first pointed out the importance of this field in a 1971 report. In a 1980 report, the Council suggested 32 important research goals and outlined promotional policies to attain them. These goals, divided into nine fields, are listed in Table 8-11.

Table 8-11. Goals for Research in Life Sciences

Field	Goal
1. Biology	Research on biological phenomena
2. Environmental science	1. Research on the relation between men and the biosphere 2. Research on physical cycles in the escosystem
3. Psychology	1. Research on functions of the brain such as perception, recognition, and behavior 2. Research on plasticity of the brain such as learning and adaptability 3. Research on neuroembryology and developmental psychology 4. Research on biological and physiological mechanism of psychopathy 5. Research on the relation between mind and body 6. Research on the relation between socio-cultural environment and mental structure
4. Health and medical care	1. Research on health, nutrition, and sports 2. Research on the aging mechanism at molecular, cellular, and organic levels 3. Research on health of mother and infant before and after delivery 4. Research on prevention, diagnosis, and treatment of congenital abnormalities 5. Research on human heredity 6. Research on prevention, diagnosis and treatment of geriatric, mental, and incurable diseases 7. Development of diagnostic and treating equipment and artificial internal organs 8. Establishment of safety evaluation techniques for chemicals 9. Research on biologically active materials 10. Research on rehabilitation

Table 8-11. continued

Field	Goal
5. Food	1. Development of technology for production and utilization of food resources 2. Research on protein and its utilization
6. Energy	1. Development of technology for production of energy using biomass 2. Research on the photosynthetic mechanism and its utilization
7. Industry applications of life sciences	1. Development of technology for industrial use of biomass 2. Development of industrial technology for living organisms
8. Population problems	1. Research on environment, resources, and population 2. Research on endocrinology and the reproductive biology
9. Recombinant DNA	1. Research on safety assessment 2. Development of a safe and useful host vector 3. Research on DNA structural analysis and synthesis 4. Research on the heterogenic manifestation mechanism 5. Development of technology for recombinant DNA production and research on micro-organisms and enzymes

Sorce: The Council for Science and Technology

At the Institute of Physical and Chemical Research of the Science and Technology Agency, research is being conducted in six of the above nine major fields proposed by the Council, with the participation of leading Japanese researchers. The Institute also performs backup work for other research laboratories, such as storage of various micro-organisms and preparation of characteristic data on animals used for experiments.

Among research projects in life science, recombinant DNA is attracting special attention because of the great contributions it is expected to make in various fields, including medicine, chemistry, and agriculture. The Science and Technology Agency, viewing research in recombinant DNA as an important base for the promotion of life science, conducts various research projects at the Institute of Physical and Chemical Research, and is planning to construct a general research laboratory with the most advanced facilities and equipment, expected to raise the overall level of recombinant DNA research in Japan.

Subjects of major special research projects in life science at national research laboratories are shown in Table 8-12.

Table 8-12. Subjects of Major Special Research Projects in Life Science at National Research Laboratories

Ministry or Agency	Research Institute	Research Subject
Science and Technology Agency	National Institute of Radiological Science	o Research on hazards of low-level radiation o Research on photosynthesis o Research on the development of a safe host vector o Development of rapidly decomposing agricultural chemicals
	Life Science Promotion Department	o Research on 6 projects including the development of artificial integral organs, bio-reactors, and utilization of new micro-organisms
Environment Agency	National Institute for Environmental Studies	o Research on the influence of air pollution on living bodies
Ministry of Education	National Research Institute for Genetics	o Research on the identification and function of genes o Research on nitrogen fixation by genetic methods
Ministry of Health and Welfare	National Institute of Public Health	o Research on liver disease and dys-hematopoiesia
	National Institute of Nutrition	o Research on prevention of hyperlipoidemia
	National Institute of Health	o Research on infection in microbial cells
	National Hygienic Laboratory	o Development of new microanalytical techniques
Ministry of Agriculture, Forestry and Fisheries	Agricultural Technic Institute, National Institute of Animal Industry, National Institute of Animal Health, Institute for Plant Virus Research, Fisheries Research Institute, and others	o Research on efficient use of natural energy in agricultural, forestry and fisheries industries o Research on efficient use of biomass o Research on nutritive mechanism of silkworm eggs o Research on operational mechanism of endocrinal control in agricultural animals o Research on pathogenic low molecular DNA o Research on cultivation of shorefish
Ministry of International Trade and Industry	Mechanical Engineering Laboratory	o Research on physical measurement technology in living bodies
	National Chemical Laboratory for Industry	o Research on chemical synthesis and utilization of nucleic acid
	Osaka Research Institute of Industrial Technology	o Research on the effective use of unused wood resources
	Nagoya Research Institute of Industrial Technology	o Research on biofunctional materials with fluorine
	Fermentation Research Institute	o Research on nitrogen fixation by micro-organisms
	Research Institute for Polymers and Textiles	o Research on high molecular materials with pharmacological effects
	National Institute for Electronical Technology	o Research on the bio-information
	Industrial Products Research Institute	o Research on compound membrane of synthetic polyamino acid
Ministry of Labor	National Institute of Industrial Health	o Research on micro-analysis of coltisol

In addition to the government-sponsored projects in Table 8-12, the pharmaceutical, food, and chemical industries have recently become active in research on recombinant DNA. However, because research on recombinant DNA makes it possible to add completely new characteristics to organisms, in 1977 the Council for Science and Technology pointed out the need for guidelines to assure safe research procedures, and in 1979 the Science Council of the Ministry of Education, after consultations with researchers in the field, issued guidelines for recombinant DNA experiments at universities. In the same year, the Council for Science and Technology issued nation-wide guidelines for such research. These guidelines, outlining basic rules for performing research on recombinant DNA, as of 1981 were revised three times by the life science committee of the Council for Science and Technology in response to changing scientific and social conditions.

(6) Disaster Prevention

Since Japan regularly experiences various natural disasters such as earthquakes, typhoons and other storms, floods, snow damage, landslides, and volcanic eruptions, understanding, controlling and preventing them is of great importance. The government, based on a report from the Council for Science and Technology, in 1981 made a basic plan for research on disaster prevention of various kinds.

Of these various projects, research on earthquake prediction, an urgent subject for Japan, is being carried out by national organizations and universities, and the results are being studied by the government-run Coordinating Committee for Earthquake Prediction. Various laboratories, such as the National Research Center for Disaster Prevention, the Civil Engineering Research Institute, and the Building Research Institute, are conducting research on the earthquake resistance of buildings and other large structures. In 1981, the Building Research Institute performed an experiment using actual-size steel-frame buildings in cooperation with the U.S.A.

Research on volcanic activities, severe weather conditions, landslides, and fires is also being actively pursued. Table 8-13 shows the budgets and activities of ministries and agencies in connection with disaster prevention.

(7) The Soft Sciences

The soft sciences develop theories, methods, and techniques to solve a wide range of problems with new analytical frames, methodologies, and techniques developed by information science, systems engineering, management science, and social science. The Council for Science and Technology, in 1971 and 1977 reports, stressed the importance of the soft sciences. In 1974 the National Institute for Research Advancement (NIRA) was established jointly by the national government, local-governments, and private companies to promote overall research on various problems facing society today.

Table 8-13. Budgets and Activities of Ministries and Agencies in Connection with Disaster Prevention

(Unit: $one thousand)

Ministry or Agency	Budget		Activity
	1980	1981	
Hokkaido Development Agency	120	116	Research on measures against damage from storms and floods
Science and Technology Agency	9,316	10,376	Research on earthquake prediction and snow damage
Environment Agency	36	36	Research on ground subsidence.
Ministry of Education	11,864	13,020	Research on earthquake, volcanic eruption, and snowslides
Ministry of Agriculture, Forestry and Fisheries	2,056	1,780	Research on protection of farm products from natural disasters
Ministry of International Trade and Industry	2,364	2,468	Research on earthquake and volcanic eruption prediction
Ministry of Transportation	7,516	5,248	Research on earthquake volcanic eruption, weather, and terrestrial and aquatic phenomena
Ministry of Posts and Telecommunications	344	832	Research on land survey and measurement technology earth quake prediction
Ministry of Labor	1,240	1,220	Research on prevention of labor disasters
Ministry of Construction	6,892	7,576	Research on crustal movements and restoration of earthquake-damaged structures
Ministry of Home Affairs	784	572	Research on prevention of fire following earthquakes and protection of oil complexes.
Total	42,532	43,244	

Source: The Science and Technology Agency

As technology begins to have a more significant impact on public life, it has become increasingly important to assess the potential effects of proposed new technology prior to its development, one of the major applications of the soft sciences. With this in mind, the government has been developing methodology for technology assessment since 1971, and as a result, technology assessment is beginning to take root.

CHAPTER 9

GOVERNMENT ASSISTANCE FOR PRIVATE RESEARCH

The Japanese government is taking active steps to promote research in the private sector, vital to the growth of the economy. This chapter will take up government assistance through subsidies, commissioning of projects, tax deductions, and special financing.

Subsidies for science and technology in 1981 amounted to $2.75 billion, up 11.5 percent from the previous year. This sum represents an increase of 90 percent, or nearly double, since 1976. A breakdown of subsidies by ministry to the private sector in 1980 and 1981, including semi-government research laboratories, is shown in Table 9-1. Of the various subsidies, some of the major ones are outlined in Table 9-2.

Table 9-1. Subsidies by Ministry

(Unit: $one million)

Ministry	1980			1981		
	A	B	Total	A	B	Total
Science and Technology Agency	1,053.2	178.3	1,231.5	1,112.6	218.6	1,331.2
Environment Agency	4.8	6.7	11.5	4.7	7.3	12.0
Ministry of Finance	–	.2	.2	–	.2	.2
Ministry of Education, Science and Culture	143.8	520.4	664.2	157.0	564.9	721.9
Ministry of Health & Welfare	41.3	12.7	54.0	45.6	14.0	59.6
Ministry of Agriculture, Forestry and Fisheries	12.7	12.0	24.7	13.0	12.0	25.0
Ministry of International Trade and Industry	133.1	343.7	476.8	143.2	440.1	583.3
Ministry of Transport	.7	2.6	3.3	13.2	2.6	15.8
Ministry of Labor	–	.2	.2	–	.2	.2
Ministry of Construction	2.1	–	2.1	2.1	–	2.1
Total	1,391.7	1,076.8	2,468.5	1,491.4	1,259.9	2,751.3

Note: A: Research expense portion of the science promotion budget and the energy-related budget
B: Research expenses other than the above

Table 9-2. Major Subsidies to the Private Sector

(unit: $ one thousand)

Ministry	Type of subsidy	Subsidy rate	Budget 1980	Budget 1981
Science and Technology Agency	Subsidies for implementation of scientific inventions (given to individuals and groups)	1/2	60	60
	Subsidies for development of uranium enrichment technology (private companies)	2/3	2,528	2,904
Ministry of Education, Science and Culture	Subsidies for publication of research results in humanities and natural sciences (researchers at universities and research laboratories)	Fixed amount	130,000	143,200
	Subsidies for promotion of research in the humanities and natural sciences (private research groups)	Fixed amount	2,028	768
	Subsidies for the purchase of research equipment (private schools)	2/3 (up to 1/2 for information processing equipment)	8,024	8,824
	Subsidies to defray operating costs at private universities	Up to 1/2	496,832	539,428
	Subsidies to defray operating costs at public medical, dental, and nursing colleges and junior colleges	Fixed amount	14,548	15,508
Ministry of Health and Welfare	Subsidies for research on food sanitation and new medicines (local governments, individuals, and private groups)	Fixed amount	4,324	4,424
	Subsidies for research on specified diseases and disabilities (local governments, individuals, and private groups)	Fixed amount 1/2	28,508	32,100
	Subsidies for cancer research (individuals)	Fixed amount	6,080	6,400
Ministry of Agriculture, Forestry and Fisheries	Subsidies for research in agriculture, forestry and fisheries (local governments, individuals, private groups)	10/10 2/3, 1/2 Fixed amount	6,452	8,196
	Subsidies for research in food industry (private groups)	Fixed amount	812	812
Ministry of International Trade and Industry	Subsidies for research in mining and manufacturing industries (private groups and individuals)	2/3 1/2	11,048	10,440
	Subsidies for research on new products and technology by small enterprises (small enterprises and local governments)	2/3 1/2 1/3	4,016	4,280
	Subsidies for research to improve technology at small enterprises (local governments)	1/2 1/4	1,884	2,012
Ministry of Transport	Subsidies for research on transportation technology (private companies)	Fixed amount	748	672

Besides these subsidies given to promote research in general, the government also assists private research in specific areas by commissioning research projects. This is particularly important in the area of high technology, which involves risks quite often too great for a private company to handle by itself, necessitating a number of government budgets for this purpose. Some of the major ones are shown in Table 9-3.

Table 9-3. Major Budgets for Commissioned Projects

(Unit $one thousand)

Ministry	Type of subsidy	Budget 1980	Budget 1981
Science and Technology Agency	Budget for research on peaceful uses of nuclear energy (private groups)	720	720
	Budget for research on radioactivity in environment and radioactivity given off by nuclear powered ships (local governments and private groups)	2,436	2,736
Environment Agency	Budget for research on environmental protection (universities and local governments)	3,912	3,852
Ministry of Health and Welfare	Budget for Japan-U.S. joint research on basic medicine	380	436
	Budget for research on pharmaceutical products (local governments and private groups)	608	768
Ministry of Agriculture, Forestry and Fisheries	Budget for research on animal breeding (local governments and private groups)	6,888	7,188
Ministry of International Trade and Industry	Budget for promotion of large-scale projects (private companies)	45,780	46,528
	Budget for research on energy saving technology (private companies)	72	424
	Budget for development of new energy (private companies)	13,052	10,600
	Budget for research on medical equipment (private companies)	3,508	3,632
	Budget for research on basic technology for future industries (private groups)	–	9,916

In addition to the above-mentioned budgets, under a separate system the government supports develpment of new technology in the private sector through such organizations as the Research Development Corporation of Japan. This corporation pays development expenses to private enterprises for technologies which are promising but difficult to develop, on condition that the company pays back the expenses if development fails. As of 1981, of the 165 projects which the corporation funded, 145, or 88 percent, were successful, while 10 were still under development and another 10 were discontinued. Recent successful projects include development of amorphous metal for electromagnetic use, a thermionic cathode made of a single lanthanum boride crystal, and an automatic pig-nursing system. A few of the recently started projects include purification of sugar liquid by magnesia, manufacture of high-strength optical thin film for lasers, and gasification of industrial plastic wastes. The corporation then helps to transfer technologies developed under this system to various industries, in line with its more general role of transferring various technologies from the government to the private sector or among private enterprises. In recent years, it has also started to transfer technologies held by Japanese companies to foreign countries. The number of cases of technology transfer totalled 164 as of 1981. The number of commissioned projects and the number of technology transfers from 1977 to 1981 are shown in Table 9-4.

Table 9-4. Number of Commissioned Projects and Technology Transfers

Item	Year	1977	1978	1979	1980	1981
Commissioned projects	Number of projects	12	11	12	11	15
	Budget for projects ($one million)	13.2	14.4	16.8	18.8	18.8
Number of technology transfers		16	25	31	30	30

The measures thus far mentioned are all forms of direct assistance by the government to promote private research. However, more indirect but equally effective measures are the granting of tax deductions and special financing.

A tax deduction system introduced in 1967 allows a company to deduct 20 percent of its annual increase in research expenses for a particular year, if that increase results in company's surpassing its highest annual research expenditure in the the past. Besides this system, tax deductions on income from technology exports, donations to public laboratories, and a depreciation allowance for assets used for research have been introduced to encourage research at private enterprises. The tax deduction for increase in research expenditures and that for income from technology exports combined are estimated to have produced a total tax savings to private enterprise of $168 million for 1981. These systems, originally scheduled to end in 1981, were extended until March, 1984, the end of the 1983 fiscal year.

Various systems of special financing for research are available, such as those offered by the Japanese Development Bank and the Smaller Enterprise Loan Corporation, which lend money at low interest rates. In 1981, the total budget for the financing system offered by the Japan Development Bank amounted to $400 million, and that of the Smaller Enterprise Loan Corporation to $20 million. Also, to assist small enterprises which have advanced technology but lack sufficient capital for its development, the Venture Enterprise Center provides financing without collateral.

CHAPTER 10

STRENGTHENING THE FOUNDATIONS OF RESEARCH

In order to promote smooth and efficient research activities in response to social and economic needs, the government is taking steps to improve the foundation for research by promoting policies in such areas as research laboratories, distribution of scientific information, and standardization, some of which will be outlined in this chapter.

(1) Research Laboratories and Personnel

Strengthening research laboratories is an essential factor in promoting efficient research activities. Various new research groups or divisions within existing laboratories are being established to meet new requirements.

Ultimately, progress in science and technology depends more on the ability and creativity of researchers than on facilities. The Ministry of Education is promoting the training of scientific personnel by strengthening higher education establishments. In addition, the Science and Technology Agency offers a program for researchers at national laboratories which enables them to further their education in Japan or abroad. The agency has also been upgrading working conditions for researchers employed at national laboratories.

(2) Distribution of Scientific Information

The Science Council of Japan, responsible for fundamental science and technology policies, since its conception has had as one of its major concerns the establishment of scientific information distribution systems. In 1969, the Council started plans to connect various informatin services through a system to be called the National Information System for Science and Technology (NIST). The NIST, made up of a number of independent sub-systems coordinated by the government to meet a variety of user needs, is based on the following goals:

(a) Promoting systematic information-gathering activities from an overall, long-term point of view
(b) Implementing policies from a national point of view with close cooperation of the relevant organizations
(c) Developing systems in compliance with the user's needs for information
(d) Attaining close coordination with other information systems
(e) Encouraging international cooperation

The structure and the roles of the NIST's member organizations are shown in Figure 10-1 and Table 10-2.

Fig. 10-1. Fundamental Structure of the NIST

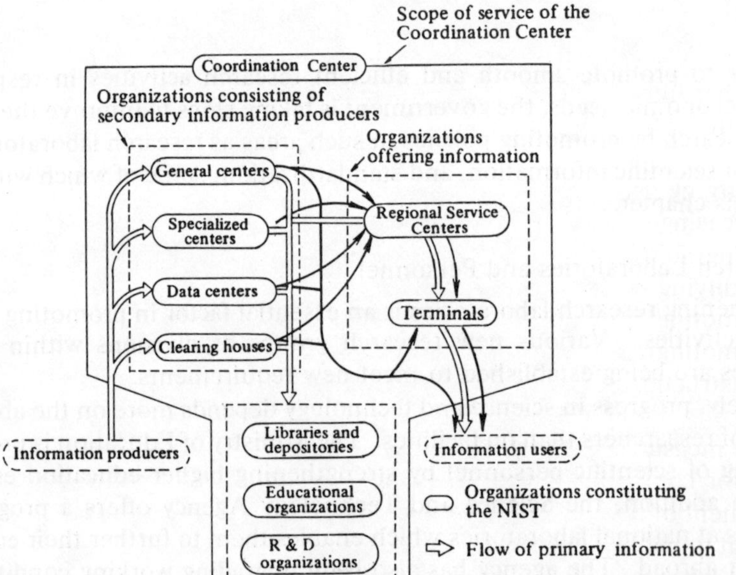

Source: The Science and Technology Agency

Table 10-2. Role of the NIST Member Organizations

Organization	Major role
Coordinating center	Promotion of the NIST activities, and coordination of the member organizations
Clearing houses	Collection, processing, and distribution of information on information sources and on-going projects
Data centers	Collection, filing, analysis, evaluation, and distribution of numerical data
Specialized centers	Collection, filing, processing, and distribution of information on specialized fields
General centers	① Collection, filing, processing, and distribution of common and basic documents ② Back-up activities of the specialized centers
Regional service centers	Information services to regional users
Libraries and dipositories	Storage and retrieval of documents
Educational organizations	Training of information specialists
Research and development organizations	Research and development of information processing technology

To further improve the activities of the NIST, the following seven goals were pointed out.

(a) Enlarging data base
(b) Increasing on-line services
(c) Strengthening primary information services and various index services
(d) Solidifying member organizations
(e) Promoting international exchange
(f) Promoting smoother operation of scientific information services
(g) Constructing pioneering systems at the Tsukuba Science City

As one measure to implement these goals, research on upgrading information services for compound materials through common networks started in 1981. Development of software systems and a data base for an on-line network is being carried out to make it possible to use information on compound materials existing at various laboratories and improve research efficiency.

A breakdown of the 1981 budget for scientific information services is shown in Table 10-3.

Table 10-3. Breakdown of 1981 Budget for Scientific Information Services

(Unit: $one thousand)

Ministry and agency	Purpose	Budget for 1981	Outline
National Diet Library		4,128	Purchase of materials related to science and technology, and automation of library services
Science Council of Japan		12	Collection of observational data
Science and Technology Agency	Promotion of the NIST	668	Strengthening of numerical data services, standardization of information, and common use of networks
	Japan Information Center of Science and Technology (JICST)	16,804	Promotion of data base, and on-line information retrieval services
	Japan Atomic Energy Research Institute	624	Processing of information on nuclear physics and engineering
	Power Reactor and Nuclear Fuel Development Corp. (PNC)	664	Collection of information on nuclear fuel resources, and fast breeder reactors
	Institute of Chemical and Physical Research	108	Collection of information on space and life sciences
	Others	1,212	
Environmental Agency	National Institute for Environmental studies	1,604	Collection and processing of information on environmental pollution
	Others	356	
Ministry of Education, Science and Culture	Budget for scientific research	696	Promotion of information science
	Publication of primary information	3,660	Publication of the results of scientific research
	Research on informatics at universities	26,496	Assistances to universities
	Information on researchers	72	Subsidy to the Japan Society for the Promotion of Science
	Others	2,092	
Ministry of Health and Welfare	Investigation on the safety of drugs	132	Collection of information on adverse reactions to medicine
	Development of medical information processing system	1,596	The Medical Information System Development Center and others
	National Hygenic Laboratory	60	Management of information activities
Ministry of Agriculture, Fisheries and Forestry	Agriculture, Forestry and Fisheries Research Council Secretariat	476	Operation of the Agriculture, Forestry and Fisheries Research Information Center and information services
	Research on the agricultural, forestry and fishery technology, and information activities	288	Subsidy to the Association for Agricultural, Forestries and Fishery Technology Information
Ministry of International Trade and Industry	Agency of Industrial Science and Technology	272	Research on data bank for physical properties and development of industrial standard information services
	Smaller Enterprise Agency	1,716	Assistance to the Small Business Information Center
	Patent Office	1,168	Preparation of abstracts of the U.S. patent specifications in Japanese
Ministry of Transport	Reinforcement of marine information management system	376	

Table 10-3. continued

Ministry and agency	Purpose	Budget for 1981	Outline
Ministry of Posts and Telecom-munications	Technology development	152	Development and investigation of data base utilization technology
	Radio Research Institute	104	Operation of ionosphere data center and purchase of associated data
Ministry of Labor	Research on hazardous substances	60	
Ministry of Construction	Public Works Research Institute	24	Budget for updating civil engineering research information system
	Total	65,620	

Note: Expenditures for development of hardware, software, and communication systems not solely designed for scientific information services are not included.

(3) Construction of Tsukuba Science City

Tsukuba Science City was established as a comprehensive research and eduational complex by transferring or newly constructing a number of national research laboratories and universities, and inviting private universities and research laboratories to join them. It is also hoped that Tsukuba Science City will help reduce population and attain balanced development in the Tokyo region. At present, the city has 45 public facilities, including national research laboratories and universities, as listed in Table 10-4. In addition, several private research laboratories have already been established in the city, and another 20 are under construction.

In order to systematically coordinate work between research and educational institutions in Tsukuba Science City, the Science and Technology Agency has established common-use facilities and the Research Institute Communication Committee with the cooperation of the relevant ministries, agencies, and research laboratories. This committee, with representatives from 53 institutions located in Tsukuba Sciency City, has established special sub-committees in the fields of library and research information, use of computers, mutual exchange of researchers, environmental safety, and public relations.

In addition, the Tsukuba Center for Institutes was set up by the Science and Technology Agency as a common facility for interchange between researchers and for quick access to science and technology information. For these purposes, the center, equipped with conference rooms, holds lectures and symposiums in support of research exchange activities.

Table 10-4. National Research Laboratories and Universities in Tsukuba Science City

Organization	Relevant ministry	Approximate area (hectares)	Transferred or constructed date
1. Educational (7 organizations)		486	
The University of Tsukuba	Ministry of Education, Science and Culture	245	1973
University of Library and Information Sciences	Ministry of Education, Science and Culture	12	1979
National Laboratory for High Energy Physics	Ministry of Education, Science and Culture	203	1971
Annex Building of the National Education Hall	Ministry of Education, Science and Culture	7	1974
Tsukuba Botanical Garden, National Science Museum	Ministry of Education, science and Culture	14	1976
Tsukuba International Center of the Japan International Coopration Agency	Ministry of Foreign Affairs	2	1979
International Agricultural Training Center of the Japan International Cooperation Agency	Ministry of Foreign Affairs	3	1981
2. Engineering (5 organizations)		213	
National Research Center for Disaster Prevention	Science and Technology Agency	27	1978
Tsukuba Telecommunication Construction Engineering Development Center, NTT	Ministry of Posts and Telecommunications	22	1975
Geographical Survey Institute	Ministry of Construction	18	1980
Public Works Research Institute of Japan	Ministry of Construction	126	1979
Building Research Institute	Ministry of Construction	20	1980
3. Science and engineering sectors (17 organizations)		302	
Tsukuba Branch of the National Research Institute for Metals	Science and Technology Agency	15	1978
National Institute for Research in Inorganic Materials	Science and Technology Agency	15	1971
Tsukuba Space Center of the National Space Development Agency of Japan (NASDA)	Science and Technology Agency	51	1972
National Institute for Environmental Studies	Environment Agency	28	1973
Tsukuba Administration Office, General Coordination Department, Agency of Industrial Science and Technology	Ministry of International Trade and Industry		1979
National Research Laboratory of Metrology	Ministry of International Trade and Industry		1979
Mechanical Engineering Laboratory	Ministry of International Trade and Industry		1979
National Chemical Laboratory for Industry	Ministry of International Trade and Industry		1979
Fermentation Research Institute	Ministry of International Trade and Industry		1979
Research Institute for Polymers and Textiles	Ministry of International Trade and Industry	140	1979

Table 10-4. continued

Organization	Relevant ministry	Approximate area (hectares)	Transferred or constructed date
Geological Survey of Japan	Ministry of International Trade and Industry		1979
Electrotechnical Laboratory	Ministry of International Trade and Industry		1979
Industrial Products Research Institute	Ministry of International Trade and Industry		1979
National Research Institute for Pollution and Resources	Ministry of International Trade and Industry		1979
Meteorological Research Institute	Ministry of Transport		1979
Aerological Observatory	Ministry of Transport	53	1974
Meteorological Observation Instrument Plant	Ministry of Transport		1974
4. Biological (15 organizations)		435	
Tsukuba Primate Center for Medical Science, National Institute for Health	Ministry of Health and Welfare	9	1978
Tsukuba Medical Plant Research Station, National Institute of Hygienic Science	Ministry of Health and Welfare	5	1979
Agricultural Technics Institute	Ministry of Agriculture, Forestry and Fisheries		1979
Upland Farming Research Center	Ministry of Agriculture, Forestry and Fisheries		1979
National Institute of Animal Industry	Ministry of Agriculture, Forestry and Fisheries		1979
Fruit Tree Research Station	Ministry of Agriculture, Forestry and Fisheries		1977
National Institute for Agricultural Engineering	Ministry of Agriculture, Forestry and Fisheries		1977
Sericultural Experiment Station	Ministry of Agriculture, Forestry and Fisheries		1979
National Research Institute for Animal Health	Ministry of Agriculture, Forestry and Fisheries		1978
National Food Research Institute	Ministry of Agriculture, Forestry and Fisheries	421	1978
Institute for Plant Virus Research	Ministry of Agriculture, Forestry and Fisheries		1977
Tropical Agriculture Research Center	Ministry of Agriculture, Forestry and Fisheries		1977
Forest and Forest Products Research Institute	Ministry of Agriculture, Forestry and Fisheries		1977
Tsukuba Office of the Secretariat of Agriculture, Forestry and Fisheries Research Council	Ministry of Agriculture, Forestry and Fisheries		1978
Tsukuba Annex Office of Seed and Seeding Division, Agricultural Production Bureau	Ministry of Agriculture, Forestry and Fisheries		1977
5. Common use facility		6	
Tsukuba Center for Institutes	Science and Technology Agency	6	1977
45 institutions in total		1,442	

Note: One hectare equals approximately 2.5 acres

(4) The International Science and Technology Exposition

To publicize science and technology and also to contribute to international exchange of science and technology, the International Science and Technology Exposition is scheduled to be held in Tsukuba Science City in 1985. This exposition, to be held in accordance with the International Exposition Treaty, is the third in Japan, following one in Osaka in 1970 and one in Okinawa in 1975.

An outline of the Tsukuba Exposition is as follows:

(a) Theme: Dwellings and Surroundings — Science and Technology for Man at Home

(b) Exposition Period: Six months, from March 17 through September 16, 1985

(c) Location: Tsukuba Science City, Ibaragi Prefecture (2 hours by train north of Tokyo)

(d) Scale:
- (a) Expected number of visitors: 20 million
- (b) Exposition site area: 102 hectares (252 acres)

(e) Expenditures:
- (a) Exposition hall construction cost: $196 million, of which approximately 44 percent will be allocated from the national treasury
- (b) Government exhibition expenditure: $146 million
- (c) Total: $342 million, of which approximately 68 percent will be allocated from the national treasury

(f) Other Related Expenditures: $1.68 billion (including improvements to connecting transportation networks)

(5) Promotion of Standardization

Standardization aims to unify terminology and drawing methods, promote technical compatibility, and rationalize productivity. National standards in Japan include the Japanese Industrial Standards (JIS), which cover a wide range of mining and manufacturing products, as the Japanese Agricultural Standards (JAS), covering agricultural, forestry and fishery products. The JIS, which have contributed to rationalization and modernization of production and distribution, as well as the spread of technology itself, at present includes about 7,700 standards covering about 1,100 products.

Along with social and economic changes in recent years, it has become vitally important to revise standards in order to improve consumer safety, conserve resources and energy, prevent environmental pollution, and encourage new technology and industries. It has also become necessary to adjust domestic standards to international ones, certify foreign products in Japan, supply information on standards and the certification system, and participate in technical standardization activities. In view of these needs, the Industrial Standardization Law was revised in April, 1980, to allow private certification organizations to inspect factories for JIS approval, and to allow foreign manufacturers to apply for JIS approval.

CHAPTER 11

TOWARD MORE CREATIVE SCIENCE AND TECHNOLOGY

Although Japan is already a world leader in such technological fields as electronics and industrial robots, the country needs to build up its power in fundamental technology as well as to direct its efforts towards such pioneering fields as space technology, life sciences, and new material engineering, which will form the core of future industry. With this in mind, it is necessary to promote science and technology for the future through close cooperation among industrial, academic, and governmental sectors in order to improve the industrial structure and the standard of living, and at the same time contribute positively to the growth and development of the world.

With the ever-increasing importance of science and technology, it is necessary to further improve existing technology while also developing creative, independent technology. Among the many issues to be resolved in the course of pursuing this goal, this chapter will take up those in research expenditures and personnel, development of innovative technology, encouragement of cooperative research, and international cooperation.

(1) Research Expenditures and Personnel

In the past 10 years, Japan's research expenditures have grown at a rate higher than those in Western Europe and the United States. The ratio of expenditures to national income, 1.96 percent in 1970, rose to 2.16 percent in 1974. Although it remained at a similar level until 1978, from 1979 onwards it started to rise again, reaching 2.42 percent in 1980, the highest ever.

But in spite of its rapid growth, this ratio is still lower than those in the United States, West Germany, and England, especially as the ratios in these countries have begun to rise in the past few years. The trend in the leading nations has been to increase research expenditures in order to revitalize their economies. France, for example, plans to increase its ratio of research expenditures to national income to 2.8 percent by 1985.

Considering the importance of science and technology in the future and the trends in other leading nations, serious effort must be made in Japan to raise research investments to 3 percent of national income, as recommended as a long-term objective by the Science and Technology Council.

To advance development of creative, independent technology, research expenditures in basic and pioneering technology must be increased, most of which need to be supported by the government. Total research expenditures in these fields, both public and private, have since 1975 been increasing at a rate of from 10 to 13 percent over each of the preceeding years, but whereas the growth rate in the private sector grew by 15 percent in 1979 and 18 percent in 1980, that of the government was 11 percent and 9 percent, respectively. After 1980, research investments by the private sector continued to be active in spite of the sluggish economy, but the growth rate of the government's research budget dropped to 8.2 percent in 1981 and then 3.5 percent in 1982. Although the economy is expected to remain sluggish, it is nevertheless desirable for the government to invest more in research for creative, independently-developed technologies.

Next, the subject of human resources will be reviewed, in particular research personnel, a major resource investment in research and development. During the past 10 years, the number of researchers in Japan has grown steadily, so that in terms of the number of researchers per 10,000 persons, the figure has grown from 16 in 1970 to 27 in 1981, second only to the U. S. among leading nations, excluding the Soviet Union. While it is important to raise the number of researchers and other research-related personnel, it is also necessary to train researchers capable of developing creative work and managing interdisciplinary projects as the need for these increases. Towards this end, it is necessary to cultivate environments, both in education and industry, that will nurture individual creativity to the fullest extent possible.

(2) Development of Innovative Technology

Creative, independently-developed technology, growing out of active fundamental research and insights into new possibilities for innovative technological applications, is indispensible to raising the level of Japan's science and technology, whose basic technology is said to be quite low. The innovative technologies referred to here consist maily of those with obvious possibilities for application in the future, once definite measures to realize them have been discovered.

At one time, it was thought that the appearance of innovative technology had slowed down globally, as suggested by a survey commissioned by the National Science Foundation of the United States, which reported that of 97 major innovations in the twenty-year period from 1953 to 1973, 32 appeared in the first five-year period from 1953 to 1958, but in the following five-year periods the number gradually dropped to 25, 24 and 16, respectively. However, in recent years, new innovative technologies have been appearing in various fields such as electronics, life sciences, and new materials, triggering significant research and development activities world-wide.

In 1981 the government established the two systems described in Chapter 2, the System for Exploratory Research for Advanced Technology, to develop creative, independent technology, and the Research and Development Project of Basic Technology for Future Industries, to develop basic technologies with major applications in the future. These two systems indicate Japan's strong awareness of the need for creative, independent technology, and promotion of research through policies of this nature will no doubt continue to be important in the future. Because developing innovative technology requires well-equipped research facilities, large investments, and long periods of research with extremely large risks beyond the capabilities of the private sector, the government must act as the core in promoting research projects. With this in mind, the Special Coordination Funds for Promotion of Science and Technology was established in 1981 in order to implement and coordinate important research activities. The 1982 appropriation emphasized development of life sciences, processes under extreme conditions, and new materials, all of which are expected to be of international importance.

As seen in Chapter 2, which describes seven areas of pioneering technology, growth in research and development in Japan following World War II has largely depended on advanced technologies from foreign nations. In the future, in addition to assuring the growth of the country's society and economy, it will also be necessary to establish independently-developed, pioneering technologies in order to make positive contributions at the international level.

(3) Encouraging Cooperative Research

Research and development in Japan is being conducted by private, academic, and public research laboratories, each carrying out a wide range of activities from basic to applied research.

However, to advance independent development of creative and pioneering technology, which requires a vast accumulation of knowledge and large investments, exchange and cooperation among private, academic, and public research laboratories is necessary.

The Science and Technology Council, having previously called for ties between these three sectors, in a 1982 report specifically pointed out the need for universities to cooperate in various areas, utilizing the results of past research. The report also noted that it is important to promote creative research and development in the private sector by, for example, government commissioning of new research projects.

Although a number of suggestions have been made from various quarters on ways of strengthening ties between industrial, academic, and public sectors, Japanese social and cultural characteristics, such as a vertically-oriented social structure and lifetime employment, have limited interaction between researchers, and to encourage creative research activities in the future, it is important to do away with such restrictions and establish a free and stimulating environment. As a step in this direction, the previously mentioned System for Exploratory Research for Advanced Technology, a system based on the individual creativity of researchers gathered from private, academic, public and foreign laboratories, was established in 1981.

In the future, it will be necessary to strengthen this type of cooperation widely over all fields of science and technology, and also to establish additional measures to link industrial, academic and public sectors.

(4) International Cooperation

With the increasing mutual dependence among various nations, international cooperation in science and technology is increasing for a variety of purposes, such as sharing research expenses, exchanging technologies, improving technological capabilities, and undertaking large-scale research projects. Japan, as one of the advanced nations of the world, is also being called upon to contribute to the solution of a variety of international issues and assist in the growth of developing nations.

In another area, as expressed at the 8th Summit Conference in June, 1982, international cooperation in science and technology is being increasingly expected to serve as a means of revitalizing the world economy. In response to this, Japan needs to further promote international cooperation in science and technology through exchange of personnel and information. Also, in order to actively take part in joint research projects, it will be necessary to strengthen the country's independent technological capabilities. Since participating in international cooperation will also result in raising the standards of domestic science and technology, it is important for Japan to actively pursue all such opportunities.